MANAGEMENT OF WORK
A SOCIO-TECHNICAL SYSTEMS APPROACH

MANAGEMENT OF WORK
A SOCIO-TECHNICAL SYSTEMS APPROACH

by
Thomas G. Cummings
University of Southern California
and
Suresh Srivastva
Case Western Reserve University

The Comparative Administration Research Institute

DISTRIBUTED BY THE KENT STATE UNIVERSITY PRESS

ISBN: 0—87338—188—2
Library of Congress Cataloging in Publication Data

Cummings, Thomas G
Management of Work.

1. Industrial management. 2. Machinery in industry. 3. Industrial organization. I. Srivastva, Suresh, 1934- joint author. II. Organization and administrative sciences. III. Title.
HD31.C84 301.18'32 76-47659
ISBN 0—87338—188—2

First printing 1977

Published by the Comparative Administration Research Institute of Kent State University
Kent, Ohio 44242

Photosetting by Thomson Press (India) Limited, New Delhi

To our families

Nancy, Sarah, and Seth
Indu, Anu, and Sahil

PREFACE

There are few topics that relate more to man's survival and growth than work. For most of us, work is a major source of economic livelihood and human enrichment. Indeed, it is difficult to imagine what life would be like without work. Given the pervasiveness of work, it is surprising that more is not known about it. What is work? Where does it fit in the context of social life? How can work be managed for greater productivity and human fulfillment? These are a few of the perennial questions that plague modern workers and managers. Answers to these queries are sorely needed if we are to manage our work lives for improved economic and social outcomes. This book is a major contribution to the management of work. It provides individuals with an understanding of work on both its social and technological dimensions; this knowledge is the basis of a strategy for managing work for increased productivity and human enrichment.

The decision to write this book grew from two related sources. First, our experiences at work as well as our informal and professional contacts with fellow workers continually raised the issue of why work was often more onerous than it should be. Put differently: Given our need to work, why wasn't it more productive and fulfilling? Attempts to answer this question continually led to the conclusion that our concepts for managing work did not do justice to our experience of work. Traditional approaches to the management of work tended to reduce work to a rationally prescribed set of activities that are carried out in formal organizations for compensation. This definition, though partially valid for many of us, seemed to limit our views of work to a narrow set of rational behavior. Intuitively, work appeared to arise and to have meaning in the nexus of interpersonal encounters that comprise social life. This led to a conceptualization of work in terms of social agreements, a view allowing us to expand work to include more of human experience. A second impetus for this book was our involvement in a long-term, field project aimed at improving the conditions and content of work. This activity led to the sobering realization that our theories of work improvement, if they are to be of use in the workplace, must meet the stringent test of practicality. This confrontation, between theory and practice, resulted in the charge that if we were really serious in our concerns for improving work, then we must make our knowledge explicit and available to those who are involved in the day-to-day management of work. This book is an attempt to meet this challenge.

The task of writing this book fell to the two of us. In many ways, however, this was only a part of the work that this material represents. Harold Bury, Rupert Chisholm, and Roger Ritvo were collaborators in the field project, and their efforts to understand and to improve work underlies much of the book. Retta Holdorf edited and typed the numerous versions of the manuscript. Her ability to manage this work was a continual source of encouragement and support. Richard Koenig from Kent State University and Ralph Kilmann from the University of Pittsburg reviewed our material and offered valuable suggestions for its improvement. Finally, Anant R. Negandhi, from The Comparative Administration Research Institute at Kent State University, supported our ideas and approach and made it possible to bring our work to the attention of a wider audience.

T.G.C.
S.S.

Cleveland, Ohio
February, 1976

CONTENTS

1

INTRODUCTION

The management of work is concerned with how people structure their relationship to technology for productive achievement. The importance of this topic has increased significantly as individuals have come to realize that their partnership with technology is not as productive or humanly fulfilling as it should be. Current problems of worker alienation, poor performance, turnover, and absenteeism are indicators of growing discontent with the conditions and content of work. Presently, there are no easy solutions to these problems. Changes on the social side of work, such as improved leadership and human relations, have not resulted in wholly satisfied or self-actualized workers. Similarly, improvements on the technical side, such as increased automation and rationalization of the production process, have not led to fully productive technologies. What is missing in many of these attempts to improve work is an integration of people and technology. This requires an approach for organizing or structuring individuals' relation to technology such that both the social and technical sides of work are more congruent. The management of work is an important step in this direction. It is a theoretical and applied strategy to managing work for both human enrichment and productivity.

The management of work is based on socio-technical systems theory and practice. This perspective offers two important advantages over current theories of work. First, it conceives of the social and technological sides of work as comprising a work system where people relate to technology for task performance. This provides an understanding of both the social and technical aspects of work as well as knowledge of the systemic interactions between the two. Second, socio-technical theory views the work system as being open to its environment. This leads to a comprehension of the interdependence between work and the wider environment. Socio-technical practice derives from these theoretical considerations. It is an approach for integrating the social and technological components of work into a unified work system, and for relating this system to its environment. The task of this book is to develop socio-technical systems theory and practice into a coherent strategy for managing work.

Like all theories of management, the management of work has its own value premises, often implicitly stated in its theory and practice. Since values influence the way we perceive reality and act upon it, it is important to explicate them as much as possible. A major value that pervades this book is the belief that work is a primary source of human enrichment. Work provides people with a structured way to master their environment, thus the very performance of work provides individuals with a powerful sense of competence or enrichment in controlling their environment. Another value is that people possess an infinite capacity to solve problems, to innovate, to learn, and to grow. The way work is managed can either enhance or thwart these capabilities. A final value that underlies this material is that collaboration between people is basic to man's existence. Individuals are social animals who live in the context of others. Since others provide man with a social identity as well as an increased capacity to act and to experience, collaboration with others is necessary for social life. Given these values, let us explore the path that this book follows.

Chapter 2 concerns historical trends in the management of work. This places the management of work in historical perspective, providing a better understanding of how people manage and hence experience work. Knowledge of the history of work is a necessary starting point for comprehending the roots of many of our current work-related problems. This provides a basis for defining work to enhance human experience, and this definition is the ground upon which socio-technical systems theory is applied.

Based on the historical review, Chapter 3 presents a definition of work derived from relationships between people. This definition leads to an understanding of work at the individual, the group, the organization, and the societal levels. It also places work in the context of human experience, where people, technology, and environment interact to form a concrete reality called work. Given this understanding of work, the remainder of the book develops a socio-technical strategy for managing it.

Chapter 4 is a brief introduction to socio-technical systems. Its origins and current applications are discussed as background for theory development.

The theoretical foundation of socio-technical systems is presented in Chapter 5. This includes an understanding of both the social and technical components of work as well as a framework for integrating them into a unified work system. The properties of this system are then discussed in terms of open systems theory.

Chapter 6 extends socio-technical theory to management practice. Two major management functions are presented: (1) relating the social and technical components to each other and (2) relating the work system to its environment.

Based on the theory and management practice previously outlined, Chapter 7 provides a strategy for applying socio-technical systems in organizations. Each stage of the strategy is discussed in terms of organizational experiments to improve work.

Chapters 8 and 9 provide concrete examples of the socio-technical strategy in action. The former chapter discusses a white collar work experiment in a forging works, while the latter presents a blue collar study in the same company.

Finally, Chapter 10 extends the socio-technical concept to the whole organization as a differentiated social system. Whereas the previous chapters were concerned primarily with the management of work—a function appropriate to systems where people relate directly to technology—this chapter involves the work of management—a function relevant to the total organization as an extended social system. Since this management function is a necessary complement to the management of work in organizations, it adds an important dimension to the socio-technical approach to management practice.

2

HISTORICAL TRENDS IN THE MANAGEMENT OF WORK

Work is one of the most prevalent realities of human experience. For the vast number of us who spend over half of our waking moments engaged in some form of employment, work seems as inevitable as taxes or death. Like other behavior essential to our existence—such as sleeping and eating—work is well known experientially, yet little understood conceptually. Ask a person to tell you about his work and he can readily recall a myriad of details about its economic, technical, and human qualities; yet ask him to define work conceptually and you are likely to hear that work is something one does in an organization for money. While this conception of work is probably true for many of us, it seems starkly bare in contrast to the richness of experience people can so easily relate about their work. In other words, people's conception of work provides few empirical clues as to why we experience work as we do.

We can seldom put into words what is essentially an ineffable quality of our experience. Although scholars have tried to capture this quality and several authors have come relatively close, there are no words that do justice to our experience of work. Indeed, we can never define conceptually the experience of work; yet people do have conceptions about work, and these conceptions structure their work lives and bestow a meaning or significance on work. In defining work, individuals set boundaries to what must be effectively managed; their definitions provide frameworks for the way work is managed. And it is the way individuals manage work that directly influences the way they experience it. Therefore, we cannot avoid the issue of defining work, for it is the starting point for the management of work. If we can develop a definition of work that places it more in the context of human experience, we will have a point of departure for managing work to enhance human experience on both its productive and satisfying dimensions.

Our approach for defining work is to examine how people have designed and managed work traditionally and then to develop a framework that combines conceptual and experiential notions. This framework for understanding work will enhance our efforts for developing strategies for planned change at the workplace. This chapter is organized in two major sections, a historical review of the concept of work and a discussion of the historical perspective and some critical issues.

HISTORICAL REVIEW OF THE CONCEPT OF WORK

Work is traditionally defined as what a person has to do to earn a livelihood. From the beginning of mankind, individuals have engaged in a multitude of activities directed at obtaining the necessities of life. Although the evolution of these activities has proceeded towards increasing levels of sophistication and ingenuity, it was not until the twentieth century that work became a topic of serious study. The advent of the Industrial Revolution was a major turning point in the study and design of work. Starting with the invention of the steam engine in the latter part of the eighteenth century and continuing into the twentieth century with the discovery that work could be scientifically analyzed and structured in a more efficient manner, the organization of work underwent a number of significant changes. Examining the way in which these changes influenced the design of work provides a clearer understanding of how people have come to manage work.

Pre-Industrial vs. Post-Industrial Revolution Period: Craft-Oriented vs. Mechanized Work

Craft-Oriented Work

Widespread during the Middle Ages, the guild or craft type of work structure was organized according to specific skill categories. Each craft was concerned with control over learning and utilization of particular skills within a tightly defined membership. Those who aspired to membership were required to serve a lengthy apprenticeship in which the art of the craft was learned from a master who was both teacher and arbitrator of craft skills. Knowledge and subsequent craft work were limited to guild members who carried out their work with a good deal of freedom and involvement.

Once an apprentice mastered the skills of his craft, he was able to exert considerable control over most facets of his work life. Depending upon his level of competence, he was able to decide upon the type and quality of goods and services to produce; he could choose his raw materials, his tools, and his methods of production; and he could market his goods and often develop new products and techniques of production. The performance of these activities enabled the craftsman to use and involve himself totally in the craft. Because of his acquired knowledge and experience, he was able to employ his sense organs to scan the work environment and to detect subtle types of information helping him guide his behavior. When certain sights, sounds, smells, feels, and even tastes were combined with know-how, intuition, and a highly trained neuro-muscular system, the craftsman was able to manipulate a few simple tools and utensils to turn raw materials into useful and often artistic products.

One need only observe a modern-day craftsman to appreciate the intricate balance and integration that exists among the different parts of his body. Each movement of his arms, hands, and fingers is coordinated to produce a purposeful change in the product; his gestures are smooth and rhythmic, yet complex and intricate in ways that are difficult to discern. The muscles that control the movement of his eyes contract and expand in ways that enable him to pick up inconspicuous visual clues; his fingers are sensitive to slight changes in pressure, heat, and texture, while his ears, nose, and tongue are keyed to sounds, smells, and tastes that are integral to the quality of his work. His facial expressions, his breathing, and his physical presence reveal a level of concentration and involvement that demands the full utilization of his whole being to produce a finished product.

The results of craft-oriented work were economically rewarding and socially and psychologically satisfying. The ability to produce a high quality service or product ensured the craftsman of economic support and community recognition. Identification with a respected and valuable service gave the craftsman a social identity that provided him with a sense of belonging and social significance. Motivation to perform depended primarily on the intrinsic rewards deriving from work that was meaningfully designed and worthy of self-respect.

In many ways, the results of craft-oriented work may be attributed to the ways in which it was "organically-designed." That is, through the skillful use of his whole self, the craftsman was able to create a work structure both adaptive to environmental demands and responsive to his own needs. The basis of this type of work design was the ability of the craftsman to regulate his behavior in the face of change. This was accomplished through a recurring process of goal-directed activity. Sensory data informed the craftsman of the consequences of his actions, and this information allowed him to create a variety of adaptive responses. By actively relating all facets of himself to his environment, the craftsman was able to modify his behavior or goals as well as influence his environment in favorable directions. This resulted in a self-regulative and inventive work structure. The results were both economically rewarding and socially and psychologically fulfilling.

Mechanized Work

The displacement of craft-oriented work began with Watt's invention of the steam engine in 1782. Once it became possible to replace man and beast as primary sources of physical work, people sought ways to mechanize their production processes. This required knowledge of what aspects of work could be performed by machines and what aspects could not. The development of relevant technology progressed throughout the next century and the knowledge and understanding of physical work that emerged from this period led to the scientific study of tasks.

Beginning with Frederick Taylor in 1911, the scientific approach to analyzing and structuring tasks became the dominant force in work design. Based on the adage that "maximum prosperity can exist only as the result of maximum productivity," Taylor and his colleagues sought ways to analyze tasks and to combine them into the most efficient method of production. One of the major outcomes of this quest was a method of work design that involved decomposing work into its simplest elementary components, specifying in detail the tasks of each component, and using the components as building blocks to produce a specified component sequence commonly referred to as a production line. Using this method, it was possible to turn production tasks into "predictable determinate mechanisms which could be performed by machines or human beings behaving like machines" (Herbst, 1966, p. 1). Given the superiority of mechanical power over muscle power, production processes were converted gradually into long sequences of steam-driven machines, each machine operating at a constant speed while performing repetitive tasks. Individuals were assigned to a limited part of the production process where they assisted machines by doing those things that a machine could not do. The logic behind this approach was both simple and persuasive: "Man was simply an extension of the machine, and obviously, the more you simplified the machine (whether its living part or its non-living part), the more you lowered costs" (Davis, 1971, p. 70).

In contrast to the industrial craftsman, the task of the production-line worker was constrained severely. Instead of having relative control over his work life, the worker was now relegated to a limited set of highly specified quasi-mechanical operations. Decisions concerning the type, quality, and amount of goods to produce, the methods of production, the acquisition of raw materials, and the distribution of products were no longer made by production workers, but rather were relegated to other specialized units. Since machines transformed raw materials into finished products, the individual's direct contact with the product was mediated through technological components. This limited the amount and type of information concerning the progress of work to that which could be inferred from the operation of a machine. It also narrowed the variety of possible adaptive responses by restricting a worker's behavior to a few simple machine adjustments geared to the design limitations of a machine. Thus, the degree of self-regulation possible was confined by a narrow range of information inputs and available responses. Since discretionary tasks were limited, the production-line worker did not have the opportunity to utilize fully his sense organs, his brain, or the rest of his body. Repetitive and highly prescribed work demanded sensory data, decisions, and behavior that were also repetitive and highly prescribed.

One need only observe a contemporary production-line worker to realize the extent to which he is not involved fully in his work. The movement of his body is smooth and rhythmic, yet limited to a few simple operations that are repeated in a relatively fixed time-cycle. His muscles flex and contract in ways that minimize wasted effort while establishing a uniform and continuous pattern. Raw materials and finished products are brought to and from his circumscribed work territory, thereby enabling him to continue his work cycle without external interruptions. Social contact is limited to his immediate work neighbors or to periodic visits from a superior who dispenses rewards and punishments. His facial expressions, his gestures, and his posture appear mesmerized by the repetitiveness of a work cycle that engages his physical self while neglecting his emotional and cognitive sides.

Judged against a criterion of economic efficiency, production-line work was extremely successful. A standardized method of production enabled workers to produce goods at a rate never before achieved. This provided social groups with the economic resources necessary to obtain a high standard of living in addition to relieving them of much of the drudgery of physical labor. Since workers were relegated to a limited part of the production process, identification with a product or service was no longer a viable means of attaining social significance and community recognition. Instead, workers acquired a social identity that placed them among the numbers of fellow men who performed similar work under similar circumstances. Individuality became submerged in work that was geared toward the masses of people who could perform standardized tasks in a prescribed manner. Because of the lack of challenge and self-control in production-line tasks, workers were left with little to which they could meaningfully relate or derive self-respect. Motivation to perform now depended on extrinsic rewards as coercion was used to elicit relevant task behavior. Failure to operate according to specification resulted in reprimands from supervisors who controlled workers' behavior and coordinated the separate parts of the production process. Thus, the workplace gradually evolved into a social hierarchy based on different gradations of discretionary power. At the bottom of this pyramid stood the worker, reduced to performing a task that was no longer under his control or worthy of his full involvement.

Like craft-oriented work, the social and economic results of production-line work were related to its design. Through a method of scientific analysis, work was decomposed into its simplest units which became the building blocks for a predictable production process. This procedure resulted in a series of tasks that can best be described as "mechanistically-designed." The requirements of each task were specified in detail so that a particular result could be obtained from a standard set of operating procedures. The basis of this form of work was a belief that maximum efficiency would result from a maximally specified task. If each worker behaved according to the one-best-method of production, predictable and efficient outcomes would follow. Variabilities that arose in the workplace—whether from the actions of workers, or the production process, or the work environment—were displaced upward to a control level that absorbed uncertainties and ensured compliance with the designed order. This blueprint operated quite well under those conditions for which it was intended, but its ability to adjust to unforeseen changes was extremely slow. Eliminating those conditions that are prerequisites for self-regulation—direct forms of information, discretionary power, and behavioral variation—forced mechanistically designed work to forfeit its adaptive capabilities in return for a standardized method of production. Workers no longer engaged their whole organism in the design of work structures responsive to environmental changes and to their own needs; instead they used limited parts of themselves in tasks scientifically designed for economic efficiency. Although the results were economically advantageous, the social and psychological benefits were often lacking.

Summary

The impact of the early stages of the Industrial Revolution on the nature and design of work is summarized in Table 1. Mechanical power replaced wind, water, and animals as the primary source of productive energy. The discovery of a predictable supply of power provided the impetus for the mechanization of the workplace. Mechanized tools and complex

Table 1

PRE-INDUSTRIAL REVOLUTION WORK VERSUS INDUSTRIAL REVOLUTION WORK

	Pre-Industrial Revolution	Industrial Revolution
Prime Examples	Craft-oriented work	Production line work
Sources of Energy	Non-mechanical power (wind, water, animals, human beings)	Mechanical power
Types of Technology	Utensils, simple tools and machines	Mechanized tools and complex machines
Type of Work Design	"Organically-designed" by workers	"Mechanistically-designed" by scientists
Task Characteristics	Variety, challenge, autonomy, direct feed-back and human contribution	Narrow, quasi-mechanical, indirect feedback and human contribution
Human Requirements	Social, cognitive, and physical parts of self	Physical parts of self
Source of Control	Self-regulation by worker	External regulation by superior
Source of Motivation	Intrinsic rewards	Extrinsic rewards
Results	Economic livelihood, socially and psycholo-gically rewarding	Economic efficiency, mass production, socially and psychologically inhibiting

machines gradually displaced utensils and tools as methods of production. Fueled by the need to discover ways to mechanize work, individuals began to study tasks scientifically. By decomposing production processes into their simplest components and specifying in detail the tasks of each component, work gradually became standardized. The advent of science in the workplace led to jobs that were mechanistically designed into a narrow set of quasi-mechanical tasks, mediated by machines, providing indirect feedback while limiting the direct contribution of the worker. Instead of fully engaging the social and cognitive parts of a worker, mechanistically designed work required a human contribution that was predominantly physical. Since tasks contained low amounts of variety, challenge, autonomy, direct feedback, and human involvement, sources of control shifted from self-regulation by workers to external regulation by supervisors. In turn, motivation to perform changed from intrinsic rewards derived from meaningful work to extrinsic rewards. The results of these radical work changes were mixed: economic efficiencies and production output soared, while the social and psychological consequences were often detrimental to workers' needs for recognition, esteem, and autonomy.

Questioning Period: The Technical Side vs. Social Side of Work: Industrial Engineering and Hawthorne Studies

Industrial Engineering and Technical Side of Work

The accumulated knowledge and understanding that emerged from the first century of the Industrial Revolution became known as industrial or production engineering. The scientific method of work design produced its greatest achievements in the mass production assembly-lines of post-World War I. The economic benefits that accrued from a standardized method of production propelled technologically advanced countries into an industrial era. Although the economic efficiency of the scientific method was clearly demonstrated, the social and psychological outcomes were frequently neglected. In practice, the design of production processes proceeded and often took precedence over the individuals who operated them. When workers were taken into account, it was usually in regard to their physiological and perceptual responses to their work environment. Although this knowledge was useful in designing machines and work areas to fit human capabilities, it did not take into account the variety of other human qualities affecting work behavior—e.g., motivation, higher-order needs, and social identification. Viewing man as an extension of the machine, industrial engineers applied a "machine theory of man" to the design of work. Man was seen as an "element or cog in a complex production system dominated by costly equipment" (Davis, 1966, p. 23). Since initiative and self-regulation were seen as possible sources of variability to the designed order of events, reward systems, supervisory control, and rigid task assignments were used to ensure specified outcomes. One of the primary consequences of this "scientific" view of man was a fundamental concern for the technological side of the enterprise, often at the expense of the social side.

Hawthorne Studies and Human Side of Work

In contrast to those scientists and engineers with a technological perspective, other researchers entered the field from a social vantage. While World War I provided a powerful impetus for mass production design principles, it also served to bring social scientists to the workplace. Focusing primarily upon individual workers, industrial psychologists developed tests for measuring individual skills, intelligence, and attitudes. They also developed procedures for measuring the physiological effects of different kinds of work. Although this knowledge was valuable for placing workers into particular jobs and determining training requirements, the social dynamics operating in the workplace were left largely unstudied. Beginning in 1924, at the height of the mass production movement, a group of social scientists under the direction of Elton Mayo began a series of inquiries that were to have a profound influence on understanding and changing the social side of work.

In what has come to be referred to as the "Hawthorne experiments," Mayo and his colleagues from Harvard University conducted a series of experiments at the Hawthorne Works of Western Electric. The primary purpose of these studies was to examine the effects of the physical environment of work on workers' behavior. Although the original intent of the study was quite orthodox in conception—the effects on production of experimentally controlled changes in rest pauses, diet, illumination, etc.—the initial results were so contrary to expectation that the researchers were forced to examine the social dynamics of the workplace in order to understand the experimental outcomes. Since these studies were to shape much of the subsequent research on the social aspects of work, a more detailed review of the experiments and of their wider consequences is needed.

The first experiment at the Hawthorne plant examined the influence of illumination on productivity. Two groups of workers were chosen for the study. One group underwent periodic fluctuations in the intensity of light, while the other served as a control group with a constant intensity of illumination. Contrary to expectations, the results showed an appreciable and almost identical increase in output for both groups. In addition to the surprise that the control group's productivity increased, the researchers observed that output rose in the experimental group even when workers experienced illumination to be worse. Since performance appeared to rise independent of the intensity of illumination, the researchers were forced to look elsewhere for an adequate explanation of the results.

Another experiment to better understand the totality of forces influencing workers' behavior in the workplace was undertaken. The Relay Assembly Test Room was chosen for experimentation. Six women who assembled telephone relays in a separate room were subject to five years of intense observation and experimentation. Changes in piece-work, rest pauses, and shorter working hours were introduced; then, these changes were gradually reduced to the original conditions, except for the retention of the piece-rate scheme. During the experiment, an observer stayed with the women and noted everything that went on. Senior company officials who were interested in the progress of the experiment frequently visited them. The results of the experiment showed a continual improvement in productivity during the experimental changes, as well as a record level of output during the period when conditions were returned to the pre-experimental state. The results were consistent with the original experiment in that performance varied independently of the experimental manipulations.

Turning to the observational data, it was discovered that an improvement in attitudes and morale had accompanied the increases in productivity. Specifically, it became evident that performance had increased because of a change in the women's attitudes to their situation. Expressions of contentment with working conditions and feelings of confidence in those in charge of the experiment were prevalent in the data. Supervision was seen as friendly and helpful, and most importantly, the women experienced themselves operating as a cohesive work group in which they helped one another for the common good of all. The major conclusions of the study were that the workers were responsive to the outside interest shown in their work by the researchers and company officials and that participation in an on-going work group provided the discipline necessary to obtain higher performance. Based on these results, the researchers shifted their focus of study from individual workers operating in isolation to informal groups of workers, their social norms, and attitudes.

Turning their attention to the social patterns of work groups, the researchers carried out an additional experiment in the Bank Wiring Observation Group. Based on the observation that groups of workers were able to exercise considerable control over the behavior of their members, the research team examined the effects of a group piece-rate system. Contrary to the previous experiments, productivity did not increase. Instead, output remained steady, even though workers were operating well within their capabilities. The results were interpreted as demonstrating the ability of group members to regulate or restrict their output in line with informal group standards—that is, members of the group applied various forms of social pressure to ensure that individual productivity did not exceed an informal group standard. The workers had formed a coherent, informal social group with its own particular attitudes, work levels, and culture. Since the group was able to control effectively the performance of its members below that expected by the company, the clash between the aims of the organization and the goals of the informal groups became obvious. The importance of the informal social group in shaping and controlling workers' behavior was not only demonstrated, but it was shown, in some circumstances, to be in direct conflict with the interests of the company.

Based on the experimental results, the research team completed their studies by carrying out an extensive interview program. More than 20,000 employees were interviewed concerning their attitudes towards themselves, their co-workers, their supervisors, and the company. The interview data corroborated the findings of the experimental studies in addition to revealing some of the actual problems faced by workers. One of the major conclusions from this research was that employee goals and behavior often followed from a "logic of sentiments" frequently in conflict with the organization's interests that followed from a "logic of efficiency." In other words, the Hawthorne Works consisted of two parallel and often opposing forms of organization: an informal one composed of informal social groups held together by an intricate web of social norms and attitudes, and a formal one comprised of highly specified work roles and task assignments held together by official company policy and an elaborate supervisory structure.

The Hawthorne experiments provided social scientists with a powerful impetus for studying the social and psychological aspects of work. Communication, leadership, interpersonal relations, and group dynamics were now viewed as important determinants of workers' behavior and attitudes. Although the enormous contribution of the Hawthorne experiments is commonly recognized, the interpretation placed on them by many social scientists and industrial managers became a serious impediment to the integration of the technological and social sides of work. Vaill (1967) summarizes these unintended consequences:

> Thus, the Hawthorne experiments supplied behavioral scientists in industry with ammunition, in a sense, with which to second-guess the efforts of industrial engineers. For many interpreters of these findings, the man's social environment had been proven more important than

> his physical environment; individual goals had been proved more important than organization goals; and a compelling slogan had been supplied: the cold ruthless "logic of efficiency" *versus* the more tender and human "logic of sentiments" (pp. 3–4).

Vaill (1967) reviews three aspects of the experiments that—contrary to the intentions of the Hawthorne investigators—helped to form a barrier between the social and technical views of work. The first aspect has to do with the finding that the attention paid to workers by supervisors and external experimenters was more important in determining behavior than the experimentally controlled changes. Vaill states: "Unfortunately, many social scientists took these findings to mean that the environment was unimportant and the social climate between superior and subordinate was all that counted. This interpretation implied that the content of a job was not worth looking at" (p. 3). A second misunderstanding involved the interview process conducted with 20,000 workers. Since the majority of the interview was aimed at workers' inner psychological responses to their job environment and not at the kinds of jobs at which they worked, many interpreters assumed that the job itself was relatively unimportant. The third way the experiments tended to place a barrier between the technological and social perspectives of work lay in Roethlisberger's and Dickson's (1939) conclusions. Based on workers' psychological responses to the organization, they concluded that the factory had been set up and run according to a "logic of efficiency" that cared only for the goals of the organization. Workers' goals and behavior followed from a "logic of sentiments" frequently in conflict with the organization's interests. Thus, the organization was assumed to be inherently in conflict with individuals' goals. In sum, the Hawthorne experiments served as an important stimulus for studying the social and psychological facets of work. At the same time, they contributed unintentionally to a segregated view of the technological and social sides of work.

Starting from a sound but often misinterpreted base of the Hawthorne experiments, social scientists focused their study of work on the tendency of the formal organization with its mechanistically-designed technology to disgruntle and dissatisfy workers. Through the 1940s and 1950s study after study showed the negative consequences when groups of workers are forced to defend themselves against a work situation unresponsive to their needs and desires. In most cases, these studies gave little attention to the technical and physical aspects of work. Instead, they opted for the man and his work group as opposed to the organization and its technology.

On the technical side of work, this era saw engineers refine their concepts about work in addition to their techniques for measuring work and setting standards. In contrast to social scientists who focused on the experiences of workers, the engineers concentrated on their performance. This dualistic approach to studying work increased the opposition between the social side of work and the technological side. In what has been referred to as the "people versus productivity" debate, both engineers and social scientists failed to respond to the relationship between people and their experiences at work and human behavior or performance. Instead of an integrated or holistic theory of work, scientists and practitioners produced a fragmented group of findings and theory.

Management Science Period: Automated Work

Starting with certain technological developments during World War II, a second phase of the Industrial Revolution began to have a major impact on the workplace. While the early stages of the Industrial Revolution were concerned with energy technology and its effect on the mechanization of work, the newer phase involved information technology and its consequence on the mechanization of decision-making. Based on the development of machines that could monitor the properties of objects and observe, generate, and manipulate symbols, scientists and engineers were able to mechanize decision-making, primarily in relation to production control. In what has come to be referred to as automation, it became possible to design machines that could monitor and control other machines. The advantages of replacing workers with automated control mechanisms were obtained by freeing production processes from much of the variability and cost of human control.

In contrast to mechanistically designed tasks, automated tasks required workers to be less physically attached to a limited part of the production process. Instead of interfacing directly with production machines, workers related to data processing machines which directly regulated production processes. The human requirements of automated tasks included greater decision-making capabilities, more discretion, less direct physical manipulation, and knowledge of a larger segment of the production process. The worker was required to be more a manager than a machine operator, more a symbol manipulator than an object manipulator, and more a planner than a doer. Automated tasks placed a premium on man's higher cognitive capabilities while relegating his physical skills to a minor position.

The development of automated technologies received its greatest impetus in the post-World War II era. Using the war-related technologies of radar, sonar, and electronic computers, scientists and engineers from a variety of disciplines sought ways to apply these new discoveries to industrial and other peace-time uses. The body of knowledge and experience that resulted from these endeavors became known as "operational research" or O.R. In the words of Russell Ackoff (1970), one of the pioneers of this approach:

> As operational research and the new technologies developed, additional fields of related studies emerged; these included information theory, decision theory, control theory, cybernetics, and general systems theory. Here, as so often had been the case in the past, "engineering" proceeded "science." Operational research workers adopted available scientific concepts, methods, techniques, and tools to their tasks, and improvised some new ones. Others were subsequently developed in the communication, decision, control and systems sciences (p. 2).

Ackoff explains the relationship of operational research to industrial engineering:

> Thus, O.R. bears the same relation to the second Industrial Revolution as industrial engineering to the first. This explains why there was so much debate in the early days of O.R. about similarities and differences. At that time the distinction between the two revolutions was not clear (p. 2).

Although the second phase of the Industrial Revolution is still in its infancy, the split between the technological and social perspectives of work continues. Operational researchers, much like industrial engineers, concentrate primarily on the technological side of work. Based on a criterion of economic efficiency, they employ various mathematical and statistical models to determine the best arrangement of the physical and financial resources of the workplace. Although this approach has done much to solve a number of crucial work problems—inventory control, resource allocation, machinery replacement, routing of goods, and sequencing of production flows—it has failed to incorporate a wide range of social factors into its methodology. Social scientists, on the other hand, have continued to focus on the social side of work while ignoring its technological aspects. Grounded on a desire to apply social science knowledge to a variety of organizational problems, social scientists have broadened their view of work to include the whole organization as a complex social system. Under the term "organization development," Bennis (1969) has defined this practice as "a complex educational strategy intended to change the beliefs, attitudes, values, and structure of organizations so that they can better adapt to new technologies, markets, and challenges" (p. 2). While the breadth of the organization-development movement has been an important step in dealing with the complex realities of work, in practice it has continued to concentrate on the social dimensions of work almost to the exclusion of the technological factors. Reviewing the history of operational research, Michael Martin (1970) concisely summarizes our argument:

> At the moment O.R. and behavioral scientists are less than enthusiastic about each other's work. The O.R. man thinks that the behavioral scientist is unable to establish verifiable and useful models of problems in industrial organizations; whilst the behavioral scientist thinks that the O.R. man is prepared to ignore or over-simplify human factors in order to build a viable mathematical model (pp. 163–164).

The movement into the second stage of the Industrial Revolution is progressing at an astonishing rate. Principles of work design that were developed for mechanized technologies—decomposition of the process into its simplest elementary components with detailed specification of tasks and of component sequences—do not readily apply to the task requirements of newer, automated technologies. Information handling, complex decision-making, planning, and process control demand a new set of work-design principles that account for the interdependencies and human capabilities needed to gather, combine, and evaluate diverse forms of information and to make expedient responses.

In addition to new work-design principles, automated forms of work also necessitate a reconceptualization of our views of man. Treating workers as if they were elements or cogs in the production process is not only an affront to the dignity of human life, but is also a serious underestimation of the human capabilities needed to operate more advanced technologies. When tasks demand high levels of vigilence, technical problem-solving skills, self-initiated behavior, and social and communication skills, it is imperative that our concepts of man be of requisite complexity. At no other period in our history has the study and design of work required a perspective that accounts for the full range of social and technological dimensions operating in the workplace. If we are to utilize and to develop technological advances and human capabilities fully, we must derive a new theory of work that incorporates the requirements of new technologies along with the social and psychological needs of workers.

Major Highlights in History of Management of Work

The history of work reveals several factors that have had a profound impact on the management of work. Starting with craft-oriented work and proceeding through the early stages of the Industrial Revolution to a more contemporary phase, the following events played a significant role in defining work.

1. Craft-oriented work presented the skilled craftsman with a great deal of discretion and autonomy. By carrying out a relatively complete task and engaging with elements of the environment, the craftsman was able to utilize himself totally. Through the skillful use of his whole organism, he was able to regulate his activities to create an organically designed work structure adaptive to environmental changes and responsive to his own needs and desires. Based on intrinsic rewards derived from work that was meaningfully designed and worthy of self-respect, the craftsman was motivated to perform high-quality work that assured him of an economic livelihood, community recognition, and social and psychological fulfillment.

2. Starting with Watt's invention of the steam engine in 1782 and continuing into the twentieth century with Taylor's discovery that work could be scientifically analyzed and restructured in a more efficient manner, work gradually became mechanized. By decomposing production processes into their simplest elementary components and specifying in detail the component tasks and sequences, it was possible to turn production tasks into predictable determinate mechanisms. Mechanized tasks presented the worker with a limited set of highly prescribed quasi-mechanical operations. Instead of having a direct relationship with his product, the worker's feedback and direct contribution were now mediated by machines. This reduced the amount of possible self-regulation by limiting the worker to a narrow set of repetitive behaviors. Motivation to perform now depended on extrinsic rewards and a coercive control structure that resulted in an organizational hierarchy based on different degrees of discretionary power. The result of this mechanistically designed form of work was a high rate of goods often produced at the expense of social and psychological rewards.

3. After World War I, the scientific study of work developed along two independent lines of inquiry. Industrial engineers employed a machine theory of man to study and design work. Concentrating primarily on the technological side of work, they derived a set of work-design principles

that resulted in a highly standardized, economically efficient form of work. Social scientists, on the other hand, focused on the social side of work. Based on the experiences of workers, they demonstrated the importance of communication, leadership, interpersonal relations, and group dynamics in determining workers' behavior. While the engineering and social science approaches did much to increase our understanding of work, there was little collaboration between the two. Instead of developing an integrated theory of work, scientists and practitioners produced a fragmented group of findings and theories which often placed the technological side of work in opposition to the social side.

4. Based on the technologies of radar, sonar, and electronic computers, those aspects of work concerned with production control gradually became mechanized. In contrast to earlier forms of mechanization based on energy technologies, the newer forms involved information technologies. Referred to as automation, the task requirements of these more advanced technologies were quite different than those of mechanized tasks. Information handling, complex decision-making, planning, and process control demanded high levels of vigilance, technical problem-solving skills, self-initiated behavior, and social and communication skills. Fractionated work designs and hierarchical forms of work organization did not readily fit these new tasks.

5. The newer phase of the Industrial Revolution is still in its infancy, and the barriers between the technological and social perspectives of work are still intact. Under the name of operational research, scientists and engineers continue to focus on the technological aspects of work. Using mathematical and statistical models, they concentrate on the best arrangement of the physical and financial resources of the workplace, almost to the exclusion of the social and psychological dimensions. On the other side of the fence, social scientists have increased their concern for applying relevant knowledge in the workplace by widening their view of work to include the whole organization as a complex social system. While the scope of this new approach—referred to as organization development—deals with the complex social realities existing in the workplace, it still does not account for the variety of technological factors affecting work behavior.

6. The movement into the newer phase of the Industrial Revolution is proceeding at an astounding rate, while our theories and practices of work design are sorely lagging. There is increasing evidence to suggest that our present view of work in addition to our theories of man are inadequate to deal with the requirements of more advanced technologies and the social and psychological needs of workers. Costly equipment and production losses combined with poor quality performance, high rates of absenteeism, turnover, and job dissatisfaction are indicators of this discrepancy. At no other time in our history does the study and design of work demand such an integration of both technological and social perspectives.

THE HISTORICAL PERSPECTIVES AND SOME CRITICAL ISSUES

The review of work design raises some fundamental issues concerning the definition of work. The various individuals who have studied and designed work have concentrated primarily on four factors: (1) the technologies employed in producing products and services, (2) the formal organizations in which much of the activities of work take place, (3) the economic incentives that accrue to individuals who perform work, and (4) the complexity in understanding the nature of work. Each of these elements has come to play an increasingly important role in defining work. Technologically, work is seen in terms of the tools, techniques, and methods of doing used for converting raw materials into finished products. Driving a truck, operating a machine, programming a computer, and working on an assembly-line are current examples of technological definitions of work. Organizationally, work has become synonomous with formal organizations. "I work for Procter and Gamble," "I am an Alcoan," and "I work for Ma Bell" exemplify this trend. Economically, work has come to be associated with paid employment. Relying on each of these factors for a definition of work confuses those elements influencing the structure and performance of work with work itself. Technological, organizational, and economic definitions of work tend to limit work to a narrowly defined set of rational activities carried out in formal organizations for compensation. The conceptual and practical problems arising from this definition have a profound impact on the way we manage and experience work.

The Consequences of Technology on Work

In the history of work design the tools, techniques, and methods of doing used in the performance of work are a dominant theme. Starting with the mechanization of simple tasks and progressing to the automation of decision-making, technology has come to play an increasingly important part in determining the conditions and organization of work. Individuals have become so accustomed to employing technological artifacts for achieving desired outcomes that work is often defined in terms of technology. When technology dominates work, people tend to experience work as a mechanical process in which machines and methods of production define the work situation. Perceiving work this way frequently leads people to invest it with a meaning or existence much like the technological objects and processes used in its performance. Work then becomes a rational, technological process existing separately from the individuals engaging in it. An unfortunate outcome of this interpretation is that human beings view work as determined by technological demands and not by their own choices—that is, people work to serve technology, and in doing this, work is under the control and direction of machines and production processes. The failure to define work as separate from technology leads individuals to experience work as being somehow disconnected from their own existence. People find it difficult to feel an integral part of production processes structured around machine-like tasks requiring only limited human involvement.

A technological definition of work overlooks the fact that work exists only through the choices and pursuits of human beings. No matter how much people are dependent upon technology for effective performance, technology does not exist independent of the social groups bringing it into existence and investing it with meaning. A definition of work embedded in technology is likely to increase human alienation from work. People will continue to experience work as separate from their own sphere of influence and involvement to the extent that they perceive work as technologically determined. Until work is seen as an integral part of the social relationships comprising human culture, a better understanding of the social conditions bringing satisfying and unsatisfying forms of work into existence is not likely to occur.

The Consequences of Human Organization on Work

The Industrial Revolution was concerned primarily with the study of work in organized settings. Starting with the scientific approach to work design and continuing to current studies of large-scale automation, the primary reference point for understanding work has been the formal organization. Based on the premise that organizations are rational instruments for structuring purposeful behavior, a major concern has been to develop principles for organizing work to achieve maximum rationality. The division of labor concept along with principles of management, such as unity of command and span of management control, are examples of this concern. A result of associating work with the traditionally defined concept of organization is that work has come to be defined as a rational set of organizationally prescribed activities for producing desired outcomes. When work is interpreted from this perspective, the rational aspects of work are often placed in opposition to the less rational human dimensions. Thus, work-related behavior is judged against a standard of rational organizational functioning. Actions not conforming to this criterion are considered unintended variance from the efficient operation of the firm, and to the extent that such deviance is disruptive of the designed order of events, individuals are placed in direct conflict with the organization.

From the Hawthorne studies to current research in organizational behavior, the clash between the formal demands of the organization and the informal and often irrationally interpreted behavior of workers has been sufficiently documented. This research leads either to the conception that work is inherently in conflict with the social and psychological needs of individuals, or that human beings are inherently lazy and irresponsible. Under the former view, work is blamed for a variety of social ills, while under the latter, workers are accused of generating organizational problems. The difficulties are not with work nor individuals per se, but with the relation between the way work is organizationally structured and the needs of workers. There is increasing evidence to suggest that this schism does not have to exist. Work can be structured to meet rational organizational requirements and individual needs. A major difficulty is the continuing belief that work is synonomous with formal organization. To the extent that this definition is used as a basis for understanding work, the clash between the rationality of the organization and the internally motivated and often irrational aspects of human behavior is likely to persist.

The Consequences of Economic Perspective on Work

A prevalent criterion for defining work is to equate it with paid employment. The majority of work studies have involved situations where individuals performed tasks for compensation. Although this focus is related to studying work in formal organizations, it raises a number of additional problems about the nature of work. Beyond the benefits of a clearly measurable yardstick for assessing work, a paid employment definition implies an economic model of man. Individuals are assumed to work for economic gain; the motivation to perform rests on money and the goods and services that it provides. Emphasizing the economic factors of work often results in the more intangible elements being either afforded a secondary importance or being excluded altogether. The personal meaning of work, the social reward of relating to others in the performance of a common task, and the cultural significance of being a productive member of society do not enter into an economic model. Excluding these dimensions from a definition of work results in a one-sided and often confusing understanding of work.

When people behave rationally to maximize their economic gain, the economic model is adequate for describing work behavior, but when they do not, the reason is difficult to explain. A good example of this problem is described in the Hawthorne studies. Much to the surprise of the researchers, groups of workers restricted production below that which could feasibly be attained despite being on a piece-rate pay system. The question of why workers would choose to earn less money was answered by examining the social sanctions and rewards of group membership. Workers were able to create and enforce group norms for what they considered was an adequate level of work. This informal standard was in direct opposition to the underlying rationale of their pay scheme which assumed no upper limit to performance other than physical capacity. Because of the social rewards of group membership and the power of group sanctions, individual workers performed at a rate which was fairly constant around the informal group standard. Thus, the ineffectiveness of purely financial incentives for maximizing production has been demonstrated empirically.

The shortcoming of a definition of work based on economic rewards rests in its inability to explain a wide variety of work-related behavior. People do work for money, but they also work for a multitude of social and psychological reasons. To the degree that wider economic factors such as low unemployment rates, wide availability of jobs, high income levels, and abundant amounts of goods and services combine to reduce workers' needs for money beyond a certain level, an economic conception of work is likely to be inadequate for understanding workers' behavior. As societies move toward these conditions, the more intangible, and in many ways, the more humanly fulfilling factors are likely to play an increasingly significant role in motivating people to engage in productive efforts.

A paid employment definition of work also limits the types of activities considered legitimate forms of work. People who perform tasks for no compensation—housewives, home gardeners, hobby enthusiasts, amateur athletes, and volunteer workers—are not perceived as engaging in work.

Although these activities are similar if not identical to those performed for compensation, a paid employment criterion fails to provide for their legitimization as work. In short, the empirical distinction between work and non-work is often not as clear as the conceptual clarity of a compensation standard. The example of a prostitute and a customer underlines this confusion. Both individuals perform similar behavior, but the prostitute receives money and is, therefore, doing work while the customer receives no money and accordingly is not performing work.

Beyond the difficulty of counting a limited set of behaviors as work, a paid employment conception leads to a number of more implicit problems. Those cultures that place a positive value on work appreciate compensated tasks more than non-compensated ones. The common belief that something done for money is more important than something done for free produces a multitude of inconsistent status differences. As an example, our neighbors experience us as valued members of the community when we are teaching classes during the academic year, but when we are home writing this book during summer vacation, they joke about the easy life of academic eggheads. Similar problems arise between husbands and wives over the value of housework versus the value of a compensated job. In those instances when a family is forced to hire a housekeeper, the distinction is more difficult to draw.

A final problem that emerges from a monetary separation of work from non-work is that researchers often study only a portion of a person's total life-space. Those who focus on work-related phenomena frequently miss a variety of other factors making a person productive and fulfilled. Family relationships, leisure activities, religious beliefs, and political affiliations have an important impact on peoples' lives. For those societies in which the time spent on leisure pursuits is rapidly increasing, the importance of studying work and non-work activities in the context of a person's total life-space is becoming a necessity. Questions about the quality of work life may soon shift towards concerns about the quality of life in general.

The Complexity of Work

The history of work raises a number of fundamental issues concerning the definition of work. Traditional studies of work concentrate primarily on technological, organizational, and economic factors for understanding work. The combination of these factors results in a definition of work as the rational activities performed in formal organizations for compensation. The problem of a definition based on these characteristics is that those elements that influence the structure and performance of work are often confused with work itself. Technological, organizational, and economic factors have an important impact on work, but they are not equivalent to work itself. Technology shapes the design of work. It provides human beings with an almost unlimited capacity to produce goods and services, and it sets constraints on the full range of human behaviors that are productively feasible. But: "No matter how much work relies on technology for effective performance, it [work] does not exist independent of the social groups that bring it into existence and bestow it with meaning."

Similarly, organizations provide a rational structure for bringing together productive resources for purposeful outcomes. To produce desired results, organizations constrain human behavior to a narrow range of rational actions. Since this often results in a clash between the requirements of organizational functioning and the needs of workers, "what is needed is a definition of work that is not in the traditionally defined concept of organization, but a definition that views work as a process of social choice among individuals engaged in common pursuits."

Finally, economic factors affect work. The availability of goods and services, the rate of unemployment, the skill level and labor rate of the work force, and costs of doing business are a few of the economic dimensions that set constraints and provide opportunities for productive achievement. The tendency to equate work with paid employment results in an emphasis on the economic factors of work almost to the exclusion of social and psychological dimensions. Yet, "people do work for money, but they also work for a multitude of social and psychological reasons." If we are to manage our work lives in a way that does justice to our experience of work, we must transcend these more abstract factors that currently define work; we must attempt to place the concept of work closer to the context of human experience.

From this perspective what can we say about defining work so that it captures more of human experience? One theme that emerges from the analysis is that work involves both technological and social choices. Those who manage work primarily from a technical perspective view work as a set of rationally determined choices that result in a maximally effective and efficient work structure. While the rationality of work cannot be denied, a definition based solely on a technological perspective often abdicates man's responsibility for the social outcomes of his technical choices. Thus, for example, we find individuals blaming people and their inherent irrationality for problems that are frequently technological. On the other hand, those who perceive work from a social perspective often limit the management of work to social choices. By placing the technical variable outside of their range of influence or choice, they frequently abdicate man's responsibility for the technical outcomes of his social choices. Then we see people blaming technology and rational organization structure for problems that are often social. Both perspectives provide important knowledge and insight about work, yet both are half-truths that must be integrated if we are to manage effectively the technological and social choices inherent in work.

A second theme arising from the history of work is that work emanates from the network of social relationships formed among human beings. Whether we are speaking of organically or mechanistically designed work, engagement in any productive activity involves relationships among people. Take the craftsman, for example: He relies on a master for training, on suppliers and buyers for a livelihood, and on peers for support and new ideas. Similarly, production-line work is the outcome of a variety of social relationships: Supervisors provide coordination and control, owners supply a place of employment and remuneration for services, and peers performing related functions bring raw materials and dispense finished products.

Because work is a product of social relationships, we often treat these relationships as the ground from which the economic, technological, and organizational figures emerge. Thus, we find that people attribute their experience of work to these more tangible factors rather than to the social relations that form the basis of work. Until we realize that work is fundamentally a social relationship, we cannot begin to manage our relations with one another for greater human enrichment.

Based on the historical trends in the management of work, we can conclude that a definition of work, if it is to account for and enhance human experience, must encompass at least two elements: (1) an integration of technological and social choices and (2) a realization that work is an important form of social relationship.

3

TOWARD A DEFINITION OF WORK

People work with other people. They depend on each other to provide the goods and services that they are not capable of providing for themselves. People also barter themselves as an instrument for providing the goods and services which others need. This inevitable interdependent relationship among human beings is a major source for organized work efforts. Work should therefore be defined in this context. Two major aspects defining it are: (1) work as an interpersonal process and (2) work as a concrete human experience.

WORK AS AN INTERPERSONAL PROCESS

Work is a social process carried out by human beings for human purposes. It originates from the network of personal relationships forming the basis for social existence. A fundamental reality of social life is that people exist in the context of others. Starting from birth and continuing throughout life, individuals must master an increasing array of environmental conditions in order to survive and to develop. One of the more pervasive features of a person's environment is other people, thus, individuals must continually strive to master their relations with others. Early in life these relationships include only a few significant others, but with maturity the variety of relations expands rapidly. Engaging with others enables people to procure biological necessities and, in time, a personal identity and the knowledge that is needed for cultural life. Work is one form of interpersonal relationship. It involves the performance of goal-directed tasks. Given the people comprising an individual's environment, work requires individuals to master a variety of interpersonal relationships. Some of these relations directly affect the task, as in the case of a group of workers performing a common task, while others indirectly affect task performance, as in the example of a wife having some influence on her husband's work life.

Like all social processes, work involves social contracts between people. Social contracts provide the structure for interpersonal relationships. They are based on mutual expectations of influence. Each person operates on the belief that he can influence the other, and to the extent to which these beliefs are shared, an interpersonal relationship is formed. Social contracts provide the structure for interpersonal relations by specifying the rules for social interaction. These rules may be formalized into an intricate body of recorded protocol or they may operate as an "unspoken law" implicitly informing people how to relate to one another. When applied to work, social contracts tie individuals to one another through mutually agreed upon tasks. What makes these relationships different from other interpersonal encounters is that the tasks have specified outcomes. In other words the social contracts of work relationships include explicit end-states. This provides work with a relatively clear boundary that is often missing in other forms of social interaction. Perhaps a simple example would clarify this point. Our decision to write this book involves a variety of relationships. One of the more salient of these relationships is our relation to our families. Because of the amount of time that it takes to write a book, we have been forced to postpone a good deal of family activity. Our families have agreed to provide us with extra time to write only until this book is finished. Therefore, we have a work relationship with our families in which there is an agreed upon task with a specified outcome. The same argument cannot be made for a variety of other relationships occurring in our lives. When we are playing with our children, for instance, we are often engaged in mutually agreed upon tasks, but unlike work, these tasks usually have no specified outcomes. We often play with our children for the sheer joy of it.

Based on the premise that work is a social process involving specific task agreements, we can formally define work as *an agreement between two or more persons to perform a stated task.* Thus, how people manage their relationships with one another is fundamental to our understanding of work. Technological, organizational, and economic factors have a profound impact on these relationships, but they are not the underlying basis of work. Work exists and has meaning only when human beings decide to relate to one another in the performance of stated tasks. Lewis Mumford (1934) in a critique of the role of technology in human culture, succinctly summarizes our argument:

> No matter how completely technics relies upon the objective procedures of the science, it does not form an independent system, like the universe: It exists as an element of human culture and it promises well or ill as the social groups that exploit it promise well or ill. The machine itself makes no demands and holds out no promises; it is the human spirit that makes demands and keeps promises (p. 7).

Similarly, no matter how much work relies on technological, organizational, and economic factors, it does not form an independent reality apart from those who create it. It exists as an element of human relationships, and it promises well or ill as those who form these relationships promise well or ill. In work, the quality of human interactions forms the basis of work that determines its productive and human outcomes.

The definition used in this book assumes that there are two major components of work: (1) an agreement between two or more persons; (2) to perform, either alone or together, a stated task. These two components help in formulating a work agreement among individuals and groups of individuals. As work increases in complexity, it requires specialization of skills, division of labor, and collaboration among people. In this sense, nothing could be accomplished unless people work together. Because of this interdependence, all work involves some level of work agreement, for the context and content of work is defined by the relevant people responsible for accomplishing the agreed upon task. These agreements are essential, since individual ability, available resources, technological, and environmental constraints require setting up a clear definition and boundary for work. The work agreement, therefore, is developed in the context of people involved in the accomplishment of work. Technological and environmental constraints also require that persons develop work agreements; for example, technology and environment tend to be forces in defining the boundary of work agreement, either through accepting the limits of technology or through development of social laws and policies for work agreements.

These issues connected with work agreements could be understood through a typology of work. The scope and boundary of the nature of work is developed on two dimensions:

1. The part of work agreements primarily determined by the person engaged in the task.
2. The part of work agreements primarily determined by other persons.

These two axis are presented in Figure 1 which illustrates the combination of these components of work agreements and helps us to derive four basic types of work agreements, thus, differentiating among four different kinds of work. The four types of work have been developed as ideal types and are named: prescribed work, contractual work, discretionary work, and emergent work.

Prescribed Work

Prescribed work assumes a hierarchical relationship among persons who develop the work agreement. Since it involves the determination of work by someone other than the person primarily responsible for the task, that person becomes a subordinate to the other. Prescribed work sets limits to the worker's use of discretion (Brown, 1965); that is, a superior sets bounds within which a subordinate may exercise his judgment. These prescriptions determine the amount of authority and responsibility that a

Figure 1

A MODEL FOR UNDERSTANDING WORK AGREEMENTS

	Determined by Self	Not Determined by Self
Determined by Others	Contractual Work	Prescribed Work
Not Determined by Others	Discretionary Work	Emergent Work

subordinate possesses; they signify those aspects of work that a subordinate is held accountable for. In the absence of clear areas of prescribed work, subordinates are unable to exercise judgment effectively. Brown (1965) summarizes this point:

> In fact, in the absence of prescribed bounds to his use of discretion, a (worker) will not know where his authority or responsibility to make decisions starts or finishes. He cannot, therefore, be held accountable either for failure to make necessary decisions or for making decisions which, in fact, usurp the authority of his own (superior). He does his work in the twilight of continuous uncertainty (p. 63).

The sanction for a work agreement for prescribed work rests in the *authority of the superior.* The worker agrees to abide by the direction of another who is in a superior position in the work relationship. Superiority does not derive necessarily from an official position, but it can emerge from a variety of other sources: expert knowledge, charisma, positive social identification, and control over rewards and punishments. Regardless of the reasons for power, the authority of the superior resides in the willingness of one person to operate within the boundaries set by another.

Contractual Work

Contractual work is determined jointly by those involved in a work relationship, the person primarily responsible for the task and some others. Contractual work involves an effort toward reduction of uncertainty. Each person reduces the uncertainty experienced by the other through the mutual exchange of commitments. This exchange may be thought of as a bargaining process in which the basis for barter is commitment. As Thompson (1967) has pointed out: "Commitments are obtained by giving commitments and uncertainty reduced for [one person] through [his] reduction of uncertainty for others" (p. 35).

The sanction for a work agreement in contractual work rests on the *authority of the contract.* Each person consents to abide by the contract and, in turn, expects the other person to do likewise. Contracts compel compliance through their enforcement by a third party or a higher authority. Legitimizing the authority of the contract, a higher authority acts as an arbitrator of the contract. Individuals in a work relationship can appeal to the judgment of this authority if they perceive that the contract is not being fulfilled. Since the right of appeal makes contractual work mutually binding, this form of work requires an effective appellate function. Each person must have a right to appeal, for without a clear appellate process, contractual work would have no basis for legitimization.

Discretionary Work

Discretionary work is directed and controlled by the person primarily responsible for the accomplishment of the task. It involves the use of judgment within limits set by contractual and prescribed work. Since this type of work is concerned with the freedom that exists in a work relationship, it is open to a wide range of variation. Some individuals fulfill their contractual obligations by merely carrying out the prescribed elements of work, while others extend their judgment to the full limits of their task. Discretionary work may be thought of as "the psychological activity of the individual in working towards and in completing the formulation of the objective" (Brown, 1965, pp. 56–57). Therefore, discretionary work goes beyond mere physical work to include the judgment and choice inherent in decision-making.

Discretionary work derives its legitimization from the *authority of the sanction*. Setting task boundaries enables superiors to define those areas of work that are under the judgment of subordinates. The sanction to use one's discretion derives from a well-defined task, for "without a clearly defined area of freedom there is no freedom" (Brown, 1965, p. 69). People are frequently required to make a variety of independent judgments in the day-to-day performance of their tasks. To the extent that they have explicit knowledge of the discretion they are authorized to use and for which they will be held responsible, the authority of the sanction legitimizes decision-making.

Emergent Work

Emergent work is environmentally determined, since it is not within the controllable boundary of either party to the work agreement. It originates from the multitude of environmental forces that operate in a work setting. Emergent work demands a response suited to the situation. Depending on the complexity and rate of change of the environment, emergent work may vary from predictable responses to recurrent conditions, to innovative responses to novel conditions. When forces in the work situation produce emergent conditions that are not controlled easily, the contractual, prescribed, and discretionary forms of work may be rendered inappropriate. Thus, contracts often need to be renegotiated with attendant modifications in the prescribed limits for discretion.

Arising from the work context, emergent work derives its sanction from the *authority of the environment*. The ability to compel action rests in the situation instead of interpersonal relationships—that is, individuals are forced to cope with the environment instead of each other. In contrast to the authorities of the superior, the contract, and the sanction, the authority of the environment derives from physical and biological laws that govern the dynamics of the work setting. Therefore, emergent work demands a response that is consistent with these laws.

Types of Work and Levels of Work Analysis

These four types of work provide a useful framework for conceptualization about work. This formulation may be applied to these levels of work analysis: the individual, the group, the organization, and the society. Examining these levels makes it possible to reconceptualize and better comprehend a number of issues currently associated with work.

Individual Work

Individuals enter into a variety of work relationships to fulfill their biological and social needs. As has been stated earlier, each work relationship provides an opportunity to engage in four types of work: prescribed, contractual, discretionary, and emergent. A primary difference between one work relationship and another is in the weighting of these types of work. Depending upon the particular persons involved in the relationship, the conditions of their work agreement, and the environmental forces influencing them, the weighting of each form of work may vary enormously. If we contrast the work of a skilled craftsman in the Middle Ages with that of a modern assembly-line worker, for example, the difference in the weighting of the four types of work is apparent.

Craft-oriented work involved a great deal of autonomy and challenge. The craftsman carried out a relatively whole task while regulating his work activities in relation to his environment. To accomplish this task the craftsman was required to perform a large proportion of discretionary and emergent work. He had control over his tools and methods of production as well as the quantity and quality of goods or services. He also encountered a variety of emergent forces as he acquired raw materials, disposed of his products, and explored new ways of working. Beyond agreements with his guild to meet specific skill requirements and contracts with customers to produce certain goods, the craftsman did relatively little contractual or prescribed types of work.

In contrast to the craftsman, a modern-day assembly-line worker performs a narrow set of quasi-mechanical operations on a limited part of a production process. His direct contribution to the product is mediated by machines and his relations to the environment are buffered by external support systems. Assembly-line work involves a high proportion of prescribed and contractual work. Individuals operate within highly prescribed boundaries that are dictated by technological requirements and division of labor work principles. To assure reliability of responses and continued work performance, voluminous contracts spell out the conditions of work relationships. Within the bounds of a highly contractual and prescribed work definition, assembly-line workers engage in limited amounts of discretionary and emergent work. These often take the form of minor task adjustments as well as deviant forms of work behavior—horseplay, output restrictions, and other types of counter-productive activity.

The primary difference between craft-oriented work and assembly-line work is in the weighting of the four types of work. The former involves large amounts of discretionary and emergent work, while the latter is weighted towards prescribed and contractual work. Although these differences may be explained by discrepancies in the people involved, the nature of their work agreements, and the environmental forces encountered, the consequences of engaging in one type of work rather than another is revealing. A craftsman, in carrying out discretionary and emergent work, was able to regulate his activities by creating a work structure adaptive to environmental changes and responsive to his own needs and desires. He was motivated to perform high quality work by intrinsic rewards deriving from work that was meaningfully designed and worthy of community recognition and self-respect. On the other hand, an assembly-line worker, in performing prescribed and contractual work, requires external supervision to regulate his activities and buffer him from environmental disturbances. Motivation to perform is contingent upon extrinsic rewards and a coercive control structure. Social and psychological rewards are often neglected as workers engage in repetitive modes of mass production.

Based on the differences between craft and assembly-line work, we can begin to draw inferences about the consequences of one type of work versus another. Since people form work relationships to fulfill biological and social needs, the forms of work that most adequately meet these needs are also most likely to be gratifying. Emergent and discretionary work demand high degrees of autonomy, ingenuity, tolerance for ambiguity, and complex decision-making. Successful task performance is likely to meet basic biological needs in addition to needs for self-control, challenge, learning, and esteem. Prescribed and contractual work require a high degree of subservience, rudimentary decision-making, and routine types of behavior. Successful task performance is likely to gratify basic biological needs as well as needs for security and certainty. Higher order social and psychological needs are not likely to be fulfilled by these forms of work. A conclusion that can be drawn is that different weightings of the four types of work inherent in a work agreement provide different opportunities for need fulfillment. Work that is loaded in favor of the prescribed and contractual elements is likely to gratify needs for biological survival, security, and certainty, while discretionary and emergent work is likely to fulfill them in addition to those for autonomy, challenge, learning, and esteem.

Because of the variations in people, there appears to be no ideal weighting of types of work that is gratifying to all workers. Depending upon the needs most salient to an individual in a work relationship, the forms of work performed can be either rewarding or dissatisfying. Current debates about the prevalence of worker alienation and job dissatisfaction can be understood in this context. Research concerning the effects of changing from prescribed work to discretionary work—that is, from limited types of work to enriched jobs—reveals that some workers are more satisfied than others (Blood and Hulin, 1967). Differences in response may be attributed to differences in individual needs. Those satisfied with enriched or discretionary forms of work are likely to possess needs for autonomy, challenge, and esteem, while the dissatisfied workers are not likely to perceive these needs as important. In short, a particular weighting of work elements that is fulfilling to one worker may not be to another. Therefore, at the individual level, a gratifying work relationship depends upon a suitable weighting of types of work that allows a worker to perform well while meeting his needs.

Group Work

A large number of work relationships involve groups of individuals who interrelate around a common task. Forming work agreements with others enables work groups to perform different types of work in a variety of environmental contexts. Groups have been found to exercise considerable control over the behavior of their members. Beyond the contractual and prescribed elements of work, workers have been shown to form informal social groups with their own work levels, attitudes, and culture. Starting with the Hawthorne studies, a considerable amount of research and first-hand knowledge has been accumulated to attest to the existence of informal groups of workers who band together to overcome or thwart the formal aspects of work. Through an assortment or clever ploys and deceptions, work groups are able to cope with excessive amounts of contractual and prescribed work by creating their own forms of discretionary and emergent work. Various types of horseplay, output restrictions, and make-work games provide workers with informal control over their work relationships in addition to increased amounts of emergent or novel conditions. The amount of energy and productive effort that goes into this behavior is an indication that workers actively seek ways to gratify those needs not met in a formal work agreement. The problem of channeling this energy into task related behavior is, in part, related to the types of work involved. When work relationships provide group members with the opportunity to engage in forms of work congruent with their needs, the power of informal group sanctions may be brought to bear on task-related problems.

The ability of work groups to take responsibility for increasing amounts of discretionary and emergent work can be seen in contemporary forms of self-regulative or autonomous work groups. These groups are provided with relatively whole tasks and clear group boundaries. Members are given the freedom to employ themselves in a variety of ways to match task requirements and changing environmental circumstances. This freedom allows group members to engage in discretionary forms of work as a response to emergent conditions. Instead of using informal group behavior to counteract formal work agreements, members apply their informal relationships towards innovative responses to task problems. Since success is predicated on the outcomes of these relationships, autonomous work groups are an attempt to elicit the energy and productive effort of informal group behavior towards task accomplishment. To the extent that workers perceive this to be a legitimate part of their work and congruent with their own needs, the informal and formal aspects of work merge.

Organizational Work

At the organization level, the hierarchical nature of work is most prominent. Weber's (1947) classical conception of bureaucracy brings to mind a pyramid-shaped structure composed of layers of work roles tied together by rules and procedures that legitimize organizational power and provide conditions for rational action. From policy makers to department heads, and downward to supervisors, foremen, and rank and file employees, the nature of work at each hierarchical level differs in the amount of time spent on each type of work. Using Parsons's (1960) notion of three levels of organizational responsibility and control—the technical, the managerial, and the institutional—we can gain a clear picture of this differential weighting of work elements.

Starting at the bottom of the organization, the technical level is involved with the effective performance of the technical task of the enterprise. Since technological functioning depends upon a high degree of rationality, this level is concerned with getting the job accomplished under conditions of certainty. The next higher level, the managerial, relates the technical level to customers and raw material resources. It also administers the technical suborganization by prescribing boundaries for task performance. The highest level, the institutional, relates the organization to its wider environment by providing higher level support and direction in the face of uncertainty. Through two-way interaction, each level reduces the uncertainty of the next lower level by prescribing task boundaries.

This process can be viewed as a series of work agreements between the three levels. The lowest type of agreement, between the technical and managerial levels, provides technical workers with high amounts of prescribed and contractual work. Contracts between managers and rank and file workers set the conditions of employment as managers articulate task-related policy. This relationship provides the certainty required for effective technical performance. The second form of work agreement, between the managerial and institutional levels, provides managers with high degrees of contractual and discretionary work. Through a process of mutual influence, managers and institutional leaders arrive at a shared agreement concerning the implementation of high-level strategy. This work contract provides a means of operationalizing the mission of the organization, and within these bounds, managers are given the discretion to set the parameters for technical performance. The highest level of work agreement is between the institutional level and the environment. Since this relationship involves elements over which the organization has little formal control, institutional leaders are required to engage in much discretionary and emergent work. Institutional leadership involves the determination of the long-term mission of the enterprise. Since it is concerned with providing the organization with a constancy of direction in the face of environmental forces, institutional leaders must choose a relevant strategy or domain while managing the emergent forces influencing the long-term survival of the firm.

Based on these differences in the weighting of the four types of work across organizational levels, it is possible to draw a number of generalizations about work in organizations. The first is concerned with the *distribution of responsibility* among the three levels of organization. Responsibility can be viewed as "the maximum time-lapse over which a person is required to exercise discretion in his work without that discretion being reviewed" (Jaques, 1965, p. 240). This time period can vary from a few hours in the case of some shop-floor workers to several years for corporate leaders. Since discretion involves the use of judgment within limits set by prescribed and contractual work, there should be varying degrees of responsibility for each level of work agreement. At the technical level, one would expect relatively low degrees of responsibility. Because of the organization's need for certainty at this level, technical workers perform within highly prescribed and contractual limits. To effectively control the technical suborganization, the managerial level is required to assess the use of discretion within these bounds. The faster a variance in performance is detected and corrected, the more certain is the technological task.

At the managerial level, the requirement of administering the technical level while operationalizing the mission of the organization results in moderate degrees of responsibility. Managers engage in contractual and discretionary forms of work to implement higher-level strategy. Since institutional leaders require information as to the effectiveness of this strategy, they monitor the use of managerial discretion at appropriate intervals. Since the consequences of implementing corporate objectives are often unknown for long periods of time, institutional leaders employ arbitrary check-points—usually in terms of accounting periods, to review the use of managerial discretion.

Finally, at the institutional level one would expect high degrees of responsibility. Relating the organization to its environment makes institutional leaders ultimately responsible for the survival of the enterprise. Although they are accountable to the owners or stockholders of the corporation, the judgment involved in performing discretionary and emergent work at this level makes evaluations of their use of discretion difficult. Decisions to enter or leave parts of the environment usually cannot be assessed in short time periods, and it may take years before commitments of the organization to future courses of action can be evaluated adequately. In place of standard time frames for assessment, the use of discretion by institutional leaders is reviewed whenever elements of the environment feel that corporate action is ineffective. Public opinion, government regulations, and informal standards of good practice are some of the controls on corporate responsibility.

A second generalization made about work in organizations involves *compensation or pay.* If responsibility varies with the type of work performed, one would expect pay to change in a similar manner. The fact that compensation increases with organizational level raises the question: What justifies these differences in pay? Based on our previous discussion, each organizational level performs different types of work. Moving from the institutional level downward, the amount of time spent on prescribed and contractual work increases or, stated in another way, the amount of choice inherent in discretionary and emergent work decreases. The need for decreasing levels of uncertainty is fundamental for the rational functioning of an organization, for without "boundaries of rationality" and the certainty that they provide, organizations would have no stable patterns of behavior for task accomplishment (March and Simon, 1958). Therefore, differences in pay are justified by the amount of uncertainty a worker is required to absorb. Institutional leaders, in performing discretionary and emergent work, assimilate high amounts of uncertainty for the organization; they provide the boundaries of rationality for the managerial level. Managers, in performing contractual and discretionary work, absorb medium degrees of uncertainty by setting the limits of rationality for the technical level. Finally, technical workers, in operating the technological task of the organization perform prescribed and contractual work that is low in uncertainty. To the extent that pay is related to uncertainty absorption, responsibility and pay are equated.

A final generalization about work in organizations involves the relationships between the four elements of work and *organizational planning.*

Since planning provides a framework for linking action to organizational objectives, the nature of the planning process is likely to vary with the type of work performed. At the technical level, organizational members are involved with the prescribed and contractual aspects of carrying out the day-to-day operations of the firm. Their behavior is concerned with the short-term goal of fulfilling current business demands. Operational plans provide the direction for getting the job accomplished as it is currently defined at this level. They set the parameters for rational task functioning by specifying the mix of resources required to produce a particular product or service. Since the technical core of an organization demands a high degree of certainty if it is to function in a rational manner, operational plans are formulated within tightly defined boundaries of rationality. At the managerial level, the focus shifts to plans for continuing in business; managers are concerned with the changing characteristics of current business. Management plans provide the framework for relating the technical core to variations in existing business. They accomplish this purpose by articulating the strategy objectives of the organization within boundaries of rationality set by the institutional level. Finally, at the institutional level, individuals are concerned with the growth and development of the firm. Discretionary and emergent types of work involve decisions about where the organization should be in the future. Strategic planning guides this process by providing the organization with a direction for getting from the present to the future. Since strategic plans provide a bridge to future environmental relationships, they go beyond existing business to include new opportunities. The uncertainties inherent in this process preclude clear boundaries of rationality, and the question of long-term organizational survival always remains unanswered.

Societal Work

The nature of work is inherently tied to the society in which it occurs. In contrast to the physical and biological environment, cultural forces are socially contrived. They exist and appear to take on a life of their own through the interaction of people who share a common meaning system that structures reality and transforms behavior into social life. Social, political, and economic factors shape work relationships and bestow them with meaning; they provide the ground from which individual, group, and organizational levels of work emerge. The way society influences work can be seen by examining the conditions that fostered the acceptance of the division of labor work design principles during the early stages of the Industrial Revolution.

The Industrial Revolution was a turning point in the organization of work. Based on the scientific study of tasks, work was structured into routine, highly prescribed tasks. The particular mix of cultural and economic conditions that existed during this era conferred a meaning on this type of work that was "supportive of a positive identity in the face of adversity" (Tausky, 1973, p. 5). The Protestant Ethic, with its belief that hard work was a road to salvation, provided a powerful source of personal identity in routine and highly prescribed work. Economic conditions of scarcity made work a personal necessity. The combination of these forces was mutually reinforcing. An ideology that valued hard work and an economy that made it a necessity were compelling reasons to engage in work regardless of its inherent content.

Today the social conditions that provided division of labor work with a certain purpose and dignity are gradually being transformed. Life styles and values affirming self-expression and freedom exist alongside traditional values of self-discipline and hard work. Although these new ideologies are still in their formative stages, the ability of the Protestant Ethic to mobilize human energy in a productive role is being called into question. The positive identity that existed in hard work irrespective of its content is slowly giving way to feelings of alienation and boredom as increasing numbers of workers experience disenchantment with traditional forms of work. Economically, an abundance of goods and services minimizes the consequences of withdrawal from work. Thus, it is becoming increasingly difficult to maintain an ethic of diligent work habits when individual and societal needs are less concerned with a scarcity of productive outputs. The accumulation of social change is altering the types of work with positive social meaning and economic validity. Prescribed and contractual forms of work, characteristic of division of labor work, are becoming untenable to increasing numbers of workers. Arguments for a shorter work week and grievances about mandatory overtime are examples of the need to withdraw from routine and highly prescribed work and to engage in other types of social relationships. Given ideologies and economic conditions fostering a withdrawal from traditional forms of work, individuals are likely to seek those conditions—often obtained in leisure pursuits—congruent with their new values and economic circumstances. If leisure is becoming a viable alternative to work, the question arises: Why is leisure more appealing than work?

Work and leisure are commonly defined as reciprocal terms: the former is associated with paid employment, and the latter is related to free or unoccupied time outside of work. Defining leisure as the converse of work leads to a number of problems that are more than merely semantic. First, the empirical difference between work and leisure is often difficult to determine. Identical activities may be seen as work or leisure depending on the person defining them or their context. The familiar example of the busman's holiday, in which one's recreation is similar if not identical to one's work, underscores this confusion. A second problem is concerned with the worth or positive meaning attributed to work rather than leisure. Since work is something one has to do to earn a livelihood, it is commonly associated with pursuits worthy of respect and personal dignity. Leisure, being envisaged as the opportunity to rest or to renew one's self to once again engage in work, is frequently related to second-class activities carried out only as an adjunct to work. The worth that we attribute to people who are gainfully employed over those who are unemployed is an example of the positive meaning associated with work rather than leisure. Third, conceptualizing work as the opposite of leisure implies that the pursuit of one negates the other; that is, leisure activities are seen as disruptive of work and vice versa. Featherbedding is an example of the negative attitudes related to mixing leisure with work, and "burning the midnight oil" is an example of the problems of combining work with leisure.

A final difficulty inherent in current definitions of work and leisure is concerned with personal identity. People gain a social identity more from their work than from their leisure pursuits. Deriving one's identity from work often negates those parts of ourselves that emerge primarily in leisure activities. Creative or artistic ability, appreciation of nature and cultural events, and special motor abilities are a few of the personal attributes usually missing from a definition of self based on work experiences. The surprise that we experience when we discover that Joe the local garage mechanic is also the president of the local ornithological society or that Mary the court stenographer also plays the flute in the local orchestra are examples of the degree to which we identify people with their work.

The confusion and problems existing in current definitions of work and leisure can be understood by examining them empirically. Work and leisure are both social processes. They are carried out in the context of others, and, therefore, they involve relationships between people. Based on an interpersonal definition, work and leisure are fundamentally the same phenomenon, although they may differ in a number of significant ways. Leisure often entails high amounts of discretionary and emergent task elements. The *Oxford English Dictionary* defines leisure as "freedom or opportunity to do something specified or implied" (1971, p. 192). Leisure activities are usually carried out in environments conducive to personal expressions of autonomy and that facilitate the exploration of novel circumstances. Indeed, many leisure pursuits are designed to allow people to use their discretion to cope with emergent environmental conditions. Athletic games, camping, boating, and bicycling are examples of people's desire to conquer emergent circumstances. Work is often weighted in the direction of prescribed and contractual elements. It is rationally designed to screen out emergent conditions that might prove disruptive to the designed order of events.

These differences, between work and leisure, are relative to the work that a person performs. Some people cannot distinguish between the two. Entertainers, professional athletes, artists, university professors, and others who perform high amounts of discretionary and emergent work often cannot differentiate work from leisure. On the other hand, those individuals who perform high amounts of prescribed and contractual work, blue collar workers, foremen, and lower-level white collar workers, have no difficulty separating the two. To the extent that work and leisure differ in terms of the weighting of task elements and in regard to environmental conditions, the conditions that drive people to withdraw from work are likely to persist. Those interpersonal relationships that provide people with discretionary and emergent tasks under suitable environmental circumstances are likely to be more appealing than those that provide routine and highly prescribed tasks in an environment suited to the regulation and control of behavior. The implications of this premise for the management of work are obvious.

Summary of Work Agreements

The nature of work as defined here places work at the core of man's social existence. Defining work as an agreement between two or more persons to perform a stated task underscores the notion that work exists only when human beings choose to relate to one another around goal-directed tasks. The interpersonal conception also stresses the central importance of the work agreement. The people involved in a work agreement and the environmental forces influencing them form four types of work which exist simultaneously in most work relationships: prescribed, contractual, discretionary, and emergent. Each of these types of work suggest a number of implications for the management of work. Prescribed work lies in the authority of the superior. It involves the direction of one person's behavior by another. The management of prescribed work demands an explicit statement of the objectives and boundaries of the task. Contractual work rests on the authority of the contract. It derives from the mutual exchange of commitments, and it requires an effective appellate process to enforce the contract and to ensure individuals the right to appeal. Discretionary work rests in the authority of the sanction. It involves the use of judgment within limits set by contractual and prescribed work. Like prescribed work, discretionary work requires a clear specification of areas of authority and responsibility. Emergent work lies in the authority of the environment. It arises from the environmental forces in the work context, and it demands a response consistent with the laws governing those forces.

The four types of work may be applied to work at the individual, the group, the organization, and the societal levels. At the individual level, work varies in the weighting of task components. Since there appears to be no ideal weighting of types of work for all individuals, a gratifying work agreement requires a configuration of work elements that enables a worker to perform well while meeting his needs. Group work also includes the four types of work. Groups have been found to generate their own forms of discretionary and emergent work to counteract excessive amounts of prescribed and contractual work. The problem of channeling this energy into task-related activities is related to the formal requirements of the task. It has been shown that relatively autonomous groups of workers are capable of using discretion to meet emergent conditions when the task allows for the legitimate use of discretionary and emergent work. Work in organizations is hierarchical. Each organizational level entails a different weighting of task components. The lowest level, the technical, is comprised predominately of prescribed and contractual forms of work. The managerial level sets the task boundaries for the technical level. It consists of large amounts of contractual and discretionary work. The highest level, the institutional, determines the long-range mission of the organization in relation to its wider environment. This task requires high amounts of discretionary and emergent work. The differential weighting of task elements among the three organizational levels also corresponds to systematic differences in responsibility, pay, and organizational planning. When viewed from a societal level, the meaning of work derives from the cultural, political, and economic conditions providing the context for work. The particular constellation of these forces at any moment in time bestows different types of work with positive social meaning and economic validity. To mobilize human energy in the performance of productive tasks, the weighting of the four types of work must be congruent with social conditions.

Performing Stated Tasks

The second part of our definition of work—the performance of a stated task—implies that work, in addition to being an interpersonal process, is a rational process. The performance of a task suggests some degree of rationality whereby people have certain cause and effect beliefs that their behavior will, in fact, lead to desired results. In respect to work, we may refer to this as "technological rationality," since the performance of a task requires an effective technology: tools, techniques, and methods of doing that enable people to produce something (Thompson, 1967). From this perspective, performing tasks requires individuals to relate to or possess certain technologies; these may be concrete objects like machines or more abstract conceptual tools such as plans or methods of work flow. Regardless of their inherent qualities, technologies enable people to transform or change their environment, through the performance of a task, into useful products or services. Thus, the performance of a task implies that work is a rational process requiring individuals to relate to technology for effective performance.

The implication that work is a rational process also suggests that it is a feedback—or goal-directed—process. When we speak of a "stated task," we mean that work has a specifically agreed upon outcome that is known to those involved in the work agreement. Thus, feedback of the results of task performance is integral to our concept of work. Without such feedback work would not be a purposeful activity for it would not be directed toward a stated goal or purpose. Feedback is necessary for any rational behavior since it guides behavior toward a desired outcome. It serves this function by providing information about the results of an individual's behavior to guide subsequent behavior; in effect, knowledge of the outcome of behavior is fed back to the person to inform him how well he is progressing toward his goal. In performing stated tasks, feedback provides knowledge that directs task performance; it tells individuals whether or not they have performed their stated tasks. Therefore, feedback is a necessary part of the rational process of work in that it provides the information necessary to determine if the work agreement has been carried out successfully.

Summary of Performing Stated Tasks

The second part of our definition of work, the performance of a stated task implies that work is a rational process. This process involves two concepts: technology and feedback. The former is required to transform the environment into a product or service, while the latter provides information on the effectiveness of task performance. Without technology or feedback, individuals would not be able to perform stated tasks.

WORK AS A CONCRETE HUMAN EXPERIENCE

Given the definition that work is an agreement between two or more persons to perform a stated task: What can we say about work as a concrete human experience? Put differently: When we speak of a concrete reality called work, what do we mean? First, work involves a specific work agreement among people. This agreement results in four types of work: pre-

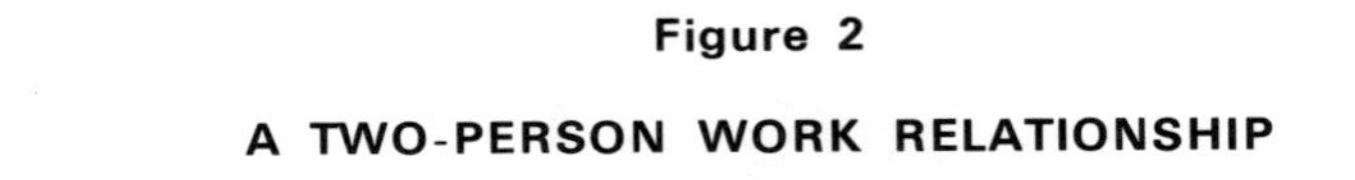

Figure 2

A TWO-PERSON WORK RELATIONSHIP

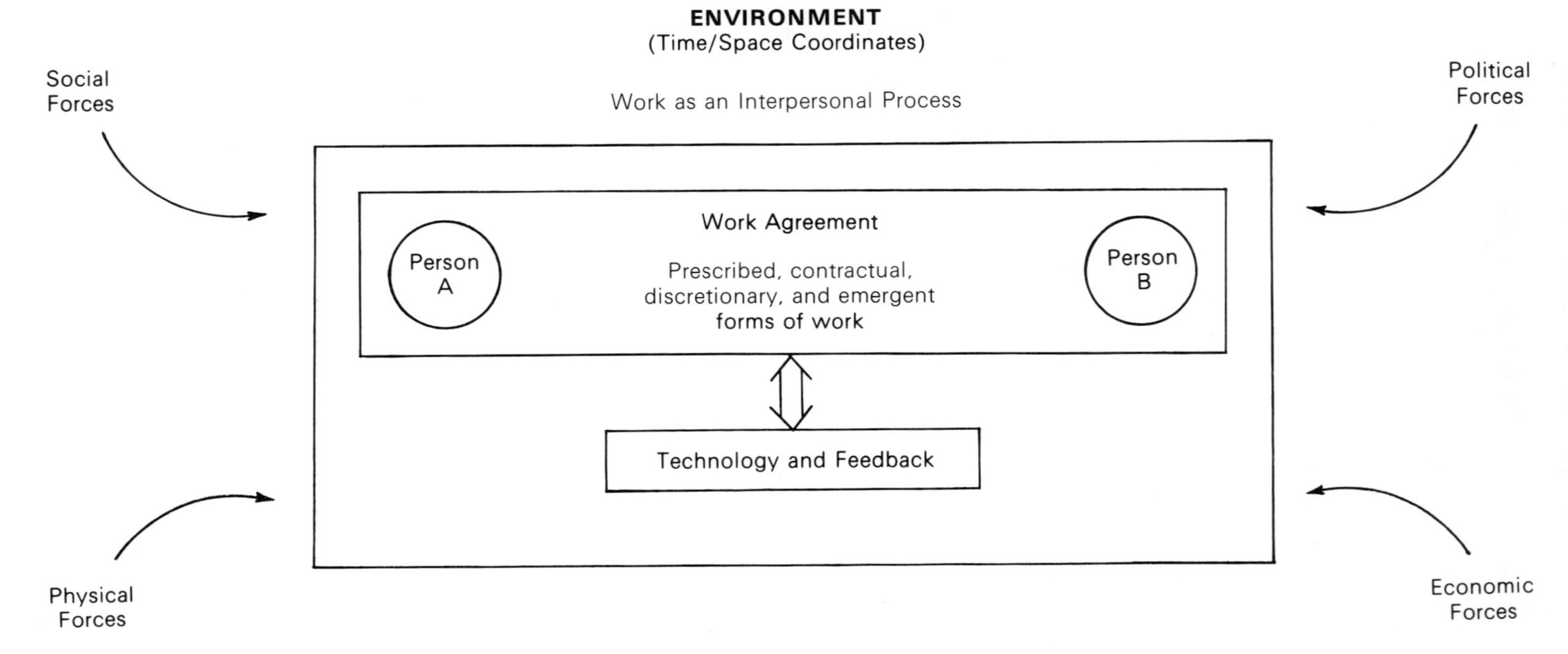

scribed, contractual, discretionary, and emergent. Second, the performance of work requires people to relate to technology to produce products or services; task performance relies on feedback of information so that the parties involved know if the work agreement is successfully executed. Finally, work takes place in a specific environment at a particular point in time and space. Thus, the economic, social, political, and physical forces present in this environment directly influence work.

Figure 2 portrays a two-person work relationship. To understand how work is experienced, we need to understand:

1. The specific aspects of the work agreement—the attributes of the people, the rules or policy of the social contract, and the weighting of the four kinds of work.
2. The technological and feedback processes operating in the work situation—characteristics of the technology and the speed, appropriateness, quantity, and quality of feedback.
3. The environmental forces operating at a particular moment in time and space—the economic, political, social, and physical forces.

The management of work requires a comprehensive framework for understanding and managing work. An ideal one would account for work at the individual, group, organizational, and societal levels. It would be congruent with our definition of work in that it would manage work to enhance human experience; that is, it would attempt to manage work agreements, technological and feedback process, and environmental forces so that individuals are both productive and enriched by their work experiences. Needless to say, there is no ideal strategy available. There is, however, a theoretical and change orientation, still in its formative stages, that is aimed in this direction. Socio-technical systems is an approach for understanding and managing work to enhance human experience on both its productive and enriching dimensions. This approach is a comprehensive attempt to integrate both the social and technical sides of work, while accounting for wider environmental forces. The remainder of this book is an attempt to present the theory and application of socio-technical systems as a basis for the management of work.

4

THE SOCIO-TECHNICAL PERSPECTIVE

BRITISH ORIGINS

The term "socio-technical systems" originates from a number of field studies carried out by Eric Trist and his colleagues at the Tavistock Institute of Human Relations in London, England (Trist and Bamforth, 1951). The socio-technical approach was a logical development from earlier field work carried out at the London factories of the Glacier Metal Works (Jaques, 1951). The Glacier studies were concerned primarily with the social dimensions of the organization. Although changes in wage and personnel policies and in a new form of worker representation helped to alleviate some of the inequities in the company, the alienation of workers remained in the form of a "split at the bottom of the executive chain" (Jaques, 1951). The researchers concluded that modifications in the social system in the absence of requisite changes in the technological system had limited consequences on the total organization.

Based on their experience at the Glacier Works, the Tavistock group reconceptualized their view of organizations in a series of field studies concerned with coal mining (Trist, et al., 1963). Because of a number of massive changes in the technology and methods of coal mining in Great Britain, an opportunity arose to study the consequences of different methods of work organization. The results of these studies provided striking evidence for the need to examine the relationship between the social and technological aspects of work as they relate to the environmental forces in the work situation. Since these studies formed the basis for much of the subsequent work in this field, a summary of this research follows.

Coal Mining Studies

The major impetus for the coal mine studies was a change from non-mechanized to mechanized mining. Previously, miners had worked under a "single place" form of mining in which pairs of workers carried out all of the required tasks with little supervision or external control. Although this approach required a large degree of physical work, miners were able to regulate their activities in relation to changes in the coal face. The advent of newer forms of mining with mechanized shovels and picks and a conveyor method of transporting coal drastically altered the single-place tradition. Designed with principles of mass-production, the newer form of mining, referred to as the "conventional long-wall method," arranged the various tasks for completing a whole cycle of coal removal into three sequential groups, covering three separate shifts. A group of workers on one shift was dependent upon the successful completion of work by the group on the previous shift and so on. Since the primary impact of this new form of work design was to isolate groups of workers performing sequential parts of the whole task, external supervision was needed to coordinate the separate groups into a whole cycle of operations.

Since coal mining is a highly variable task that differs with the conditions of the coal face, it is imperative that workers regulate the amount of time spent on various sequential tasks. The absence of this control may result in one set of tasks impeding the performance of a subsequent set of tasks. (The conventional long-wall method mitigated this problem.) Workers from one shift had difficulty regulating their work in relation to the work of other shifts. External supervision was unable to remedy this situation, since the conditions of underground mining made it difficult to control and to coordinate workers effectively. The inability to cope with the requirements of the task resulted in hostility and conflict among the workers, and between workers and management. The observable consequences of this problem were low productivity, high turnover and absenteeism, and various forms of accidents and psychologically related problems. In sum, a mass-production form of work design, developed primarily for standardized forms of work, did not fit the variable task conditions of a changing underground environment. The normal tensions existing in an accident-prone, underground situation served to amplify the consequences of this mismatch in a rather dramatic fashion.

Against this background, researchers from the Tavistock Institute studied a number of natural field experiments in which groups of workers tried new forms of work designs aimed at alleviating the problems inherent in the conventional long-wall method. Referred to as the "composite long-wall method," relatively autonomous groups of workers took responsibility for a whole cycle of mining operations. Based on a common pay rate, the men were able to interchange their tasks depending upon changes at the coal face. Although each of these composite groups developed unique ways of dealing with the underground situation, the results of this more adaptive method were positive. Productivity rose as the groups were able to complete effectively a whole cycle of mining tasks; labor turnover, absenteeism, and accident rates declined as the men experienced less tension and conflict-inducing conditions than before. It is interesting to note that the composite form of work design resembled the earlier method of single-place working employed in non-mechanized mining; engineers and managers had failed to recognize its potential applicability under the newer mechanized conditions.

The coal mine studies provided two major breakthroughs for the study and design of work. First, they demonstrated empirically the validity of viewing work structures as open, socio-technical systems. The intricate and often subtle relationships between the psycho-social and the technological dimensions of the coal face demonstrated the necessity of examining this critical interface in relation to environmental forces. Trist and Bamforth (1951) state: "So close is the relationship between the various aspects that the social and psychological can be understood only in terms of the detailed engineering facts and of the way the technological system as a whole behaves in the environment of the underground situation" (p. 11). Second, the research provided a viable alternative to mass-production work design principles in the form of composite or autonomous group concepts. Although the investigators did not advocate the use of autonomous work groups in all productive settings, their findings showed clearly that more than one form of work organization can effectively operate the same technology. They concluded that the superiority of one work organization over another lies in the extent to which it meets the requirements of the task and the social and psychological needs of the workers. In other words, the coal mine studies demonstrated that there is an element of organizational choice in designing work.

Weaving Studies

Based on the early findings of the coal mine studies, A. K. Rice (1958), from the Tavistock Institute, applied a socio-technical systems approach to the analysis and redesign of both non-automatic and automatic weaving sheds in Ahmedabad, India. Rice's analysis concentrated primarily on worker-machine assignments, task groupings, and supervisory roles. The outcome of his analysis revealed an organizational structure that hindered interaction among workers performing highly interdependent tasks. This resulted in problems of coordination and continuity in production. With the help of workers and managers, Rice was able to redesign the work structure so that interdependent workers were made part of relatively autonomous work teams that were responsible for the cperation and maintenance of a group of looms. Supervisors were then provided with unified commands that corresponded to the natural task groupings. The results of the redesign were increased productive efficiency and reduced thread breakage.

The Ahmedabad experiments were a powerful confirmation of the socio-technical approach. First, they added to the general applicability of the socio-technical perspective, since they were carried out in a different cultural setting than the coal mine studies and they included a technology quite different from coal mining. Second, the weaving studies demonstrated that a socio-technical approach could be used to intervene actively in the on-going operations of an organization. The coal mine studies, on the other hand, were concerned primarily with natural field experiments which unfolded in the normal course of coal mining. Third, Rice's studies included a complete reorganization of the managerial structure of the company. The importance of looking beyond the immediate work setting to larger segments of the organization, including the managerial structure, was confirmed. Fourth, the success of the autonomous group structure reaffirmed the notion or organizational choice and contributed to the general applicability of autonomous work principles.

AMERICAN ORIGINS

Socio-technical systems is commonly associated with the pioneering work of the Tavistock researchers, although a parallel development in the United States produced similar results. Under the term "job design," Louis Davis (1957a) and his colleagues at the University of California at Berkeley carried out a series of field experiments similar to the Tavistock studies. Starting from a background in industrial engineering, the Berkeley group examined the relationship between task factors, social and psychological needs of workers, and organizational processes. A major outcome of these studies was the development of a "job-centered approach" to work design. This approach was based on the premise that jobs involve two sets of variables: (1) those introduced by the technical process, the organization, and the person, and (2) those that stem from the interaction of these elements. Since the interaction of these elements is crucial to the effective design of jobs, they concluded that job design must jointly account for the needs of the organization, the technological process, and the worker. The major contribution of the job design studies was to widen the focus of work design to include relevant organizational variables and social and psychological needs of workers in addition to the more traditional concerns for technological factors and the physical and perceptual characteristics of workers. The fact that this approach was developed by engineers—rather than social scientists—made its impact on work design all the more powerful.

Since the advent of the socio-technical systems approach in the early 1950s, similar experiments have been conducted in a variety of countries, technological settings, and work populations. Their results have contributed to the systematic development of the theory and application of the socio-technical perspective. During the course of this development, Davis and Trist began to collaborate in this effort, and in 1967 they started the first graduate program in socio-technical systems at the Graduate Business School at the University of California at Los Angeles. Recent studies have been concerned with the wider impact of work on the quality of life in post-industrial societies. The Norwegian Industrial Democracy Project is an example of this current interest in large-scale problems.

INDUSTRIAL DEMOCRACY

The Industrial Democracy Project began in 1961 as a response to increasing demands from Norwegian workers for representation on boards of management. To learn the extent to which such representation leads to decreases in alienation and increases in productivity, the Norwegian Confederation of Employers and the Norwegian Confederation of Labor jointly sponsored relevant research. The research was conducted by Einar Thorsurd and his co-workers at the Institute of Social Research in Trondheim, Norway. Researchers from the Tavistock Institute of Human Relations also collaborated.

The first phase (Phase A) of the project consisted of a series of interviews at five Norwegian firms that had first-hand experience with employee representation on managerial boards. The results showed that little happened when workers were added to managerial boards—rank and file participation, alienation, and productivity remained relatively unaffected. These findings were consistent with similar experiences in other countries: workers' councils in Yugoslavia, West Germany, and Great Britain, and more direct forms of participation such as human relations programs in Norway, the Netherlands, and the United States (Van Bienum and De Bel, 1968).

The primary conclusion from this phase of the project was that workers' representation on managerial boards did not provide individual employees with direct participation in the day-to-day operations of the enterprise. Emery and Thorsurd (1969) summarize this conclusion: "Briefly, what we are suggesting is that two of the necessary conditions for the emergence of a higher level of participation are not present: These are that the individual should have more elbow-room within his job, and second, greater responsibility for decisions affecting his job" (p. 30).

The results of Phase A of the project formed the basis for a second phase (Phase B) concerned with finding ways for more direct, personal participation in the workplace. In short, industrial democracy was reformulated to mean: direct participation in the daily operations of the enterprise. To determine the conditions for personal participation in the workplace, a series of socio-technical experiments was carried out in selected plants in key industries in Norway. Since the results of these experiments were to be evaluated and disseminated to other sectors of the Norwegian economy, it was necessary to select organizations which both supported experimentation and possessed national influence and prestige. Towards this end, members from both confederations and sector committees from the industries concerned selected the experimental sites and sanctioned the experiments. It was hoped that sanctioning from a national level would provide the influence necessary to allow for a wider diffusion of results into the Norwegian society.

In 1963 the first experiment was carried out in the metal-working industry. Subsequent experiments were conducted in organizations in the pulp and paper, chemical, and shipping industries. Each experiment involved modifications of existing work practices under conditions of voluntary participation by management and workers. Great care was taken to engage relevant organizational members in each phase of the experiments. Although the experiments were carried out in highly diverse settings, the results confirmed the general hypothesis that individual participation in the decisions and methods of carrying out the day-to-day tasks of the organization leads to reduced alienation and increased productivity.

The industrial democracy project is currently in its dissemination phase. Relevant knowledge is being diffused by various means, and the results of this process are under evaluation. Although the overall impact of the project is not fully known, preliminary findings are optimistic. One of the interesting outcomes of the project is the apparent spill-over of active participation from the shop floor into wider aspects of the Norwegian society. Although the data are still incomplete, it appears that participation in the workplace may lead to more active participation in other sectors of a person's life. Indeed, Norway is rapidly becoming a living example that a new concept of industrial democracy, based on socio-technical systems principles, is not only viable for organizational life, but also facilitates democratic ideals in the wider society.

SUMMARY

Socio-technical systems is a comprehensive framework for understanding and managing work. The theoretical and applied direction of this approach is consistent with an interpersonal definition of work. The premise that work is composed of both social and technological systems is congruent with the concept of a work agreement and of its relationship to technology. The fact that work requires individuals to master technological processes makes the interface between the social and the technical systems central to an understanding of work. The notion that a socio-technical system relates to an environment that both influences and is influenced by the work system is compatible with the idea that a work agreement, a technology, and an environment interact to form a concrete experience called work.

5

A THEORETICAL FOUNDATION FOR UNDERSTANDING SOCIO-TECHNICAL SYSTEMS

Socio-technical theory is grounded on two fundamental premises: (1) that the production of a good or service requires the joint operation of two independent, yet correlative systems—a social and a technical system; (2) that the social-plus-technical system must relate to its environment if it is to function and develop. To understand the first premise, both the social and technical systems have to be examined as independent elements of work and then placed in a framework for relating them into a jointly operating socio-technical system. The second premise, relating the socio-technical system to its environment, starts from a definition of socio-technical systems that differentiates them for their environment; then proceeds to open system concepts for relating them to their environments.

THE SOCIAL SYSTEM

Social system can be described from many vantage points. In the present context, the term "social system" will be used to connote a relationship between people who interact with each other in a given environment for the basic purpose of achieving an agreed upon task or goal. The nature of the task or goal could be as varied as the people participating in it and the environment in which the task is to be accomplished. This task or goal also requires that people develop mechanisms for its accomplishment and these mechanisms are basic components of the social system. For the present purposes it is adequate to discuss two major human processes which enhance and maintain a social system of a workplace: interaction among human beings and human exchange with the environment.

Human Interaction

Talking, dancing, drama and work are a few examples of the forms of human interaction. Some are clearly purposeful, while in others the purpose might be more apparent than real. But all human interactions have consequence on the participants. Human interaction serves the purpose of observing, evaluating, and changing the consequences of such interaction among its participants. Specifically, human interaction serves the following purposes:

1. As mechanisms of human influence on human beings: It is through dialogue, physical contact, conversation, engagement with each other that human beings influence each other. From birth to death human beings

are engaged with each other in loving, fighting, helping, building, destroying, and training future generations to learn similar behaviors, attitudes, and skills. In all its forms, the transmission of life-style to others is an expression of human influence.

2. For the development of common values: A sustaining part of society is the presence of some common values. These are nurtured by and evolve through human interaction. It is a commonplace experience to notice the evolution of common values as a part of working towards an agreed upon goal. The development of common goals and values, for example, is a prerequisite for accomplishing any task which needs the participation of more that one person.

3. As mechanisms for social and group integration: People need to maintain a level of integration in their social relationship and continue to evolve language, signs, and symbols through which human interaction is facilitated. The notion of social interaction is a by-product of the development of such instruments as language, signs, and symbols. Observation of different societies indicates that common language, signs, and symbols act as social integrators, and social integration is possible only through human interaction using common instruments. At the workplace, people working together develop their own group language for machines, men, and events. Superior-subordinate, friendship, union-management, departmental relationships, and the like, are some examples of human relationship at the workplace. Such relationships are maintained through the processes of interaction. The nature of these relationships also affects the ability to solve problems and manage conflicts in the context of accomplishing any task.

Human Exchange

Human beings are constantly in a process of exchange with the environment. We breathe the air, we eat the products of the earth, we handle tools and machines, and we constantly create new environments through human exchange. Such exchange with environment serves the following purposes:

1. It is a mechanism for manipulation and control of the environment: Beginning with birth, human beings are constantly involved in accommodating the environment and the objects in the environment. We learn to play with the toys instead of chewing them, we learn to climb out of the crib instead of falling from it, we learn to decorate our surroundings instead of leaving them dirty and unattended. These experiences are extended in later parts of life in terms of our development of multiple skills for manipulating objects like automobiles and, machines.

2. It is a mechanism for the creation of new environments. The changing nature of human exchange is embedded in the notion of changing environments and its corollary, the human intervention in the environment. We change environments through needing speedier automobiles and airplanes. The noise level created by jumbo jets, for example, is unacceptable to some human beings, and thus, those affected by it have begun demanding a change. These demands are simple seeds for the creation of a new environment consistent with the needs for human exchange.

3. It is a mechanism for the development of new technology. As speedier cars and airplanes create damage to the environment through air and noise pollution, the need for new technology is felt through human exchange. The consequences of pollution, for example, experienced by human beings, require that we invent technologies that are helpful for human exchange instead of detrimental for it.

The nature of the social system explained here suggests that technological development is a consequence of the human process in the social system and that, in turn, technology becomes one of the major components in the development and maintenance of the social system. Figure 1 illustrates the human interaction and human exchange processes that comprise the social system of work.

THE TECHNOLOGICAL SYSTEM

The technological system consists of the tools, techniques, and methods of doing that are employed for task performance. Unlike the social system, which is a self-generating system, the technological system cannot generate itself; it exists and has meaning only when social groups bring it into existence and bestow it with meaning. In this sense, the technological system is an artifact, created by human beings to serve human purposes. In conceiving of technology, we are often conditioned to think in terms of concrete objects: tools, machines, and gadgets. Since these objects are readily available to our sensory receptors, they can be easily seen, heard, smelled, felt, and even tasted. Indeed, we are so accustomed to relying on our senses for knowledge of the environment that we frequently limit our

Figure 1

DIAGRAMMATIC REPRESENTATION OF PROCESSES AND CHARACTERISTICS OF A SOCIAL SYSTEM

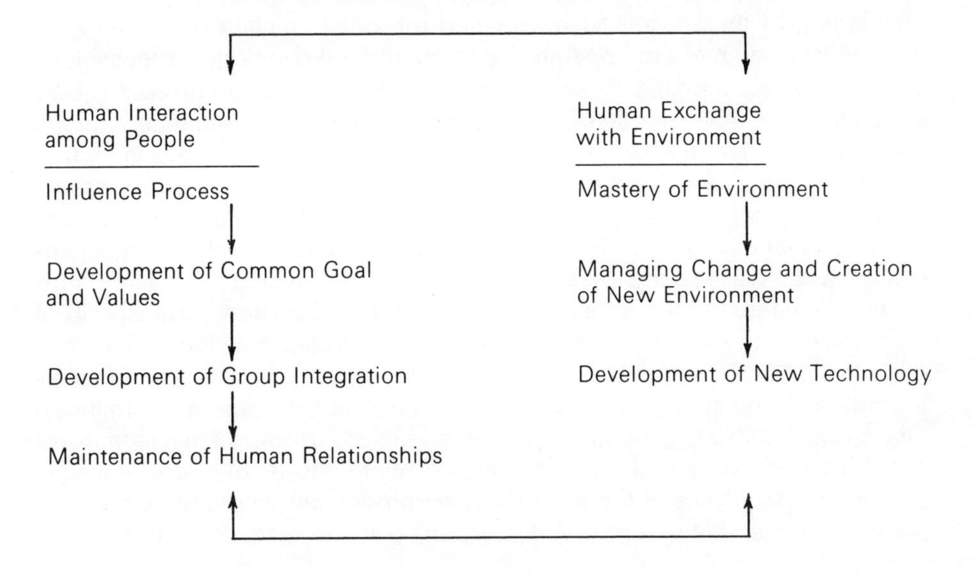

views of technology to those things with direct impact on our senses. If we widen our views of technology to include those types of knowledge helping to organize and direct our productive efforts, the concept of technology takes on a wider perspective. Ideas, procedures, and methods of production are some of these more abstract forms of technology. Although many of these latter kinds of technologies reside in books or in other written records and often in the minds of men, nonetheless, they are integral parts of our technological heritage.

Although the technological system does not have an existence apart from the social groups that create it and attribute a significance to its use, it does operate according to a set of laws that differ from those of the social system. In contrast to the social system that functions according to laws of the animate—biological and psycho-social laws—the technological system operates according to laws of the inanimate—mechanics, hydraulics, electronics. Both kinds of laws operate jointly when social groups employ technology for task accomplishment; yet both systems may be treated as independent by virtue of the different kinds of laws that govern their behavior. When we talk of a social system, we are referring to the people and the relationships between them that are the basis for a task agreement; when we refer to a technological system, we mean the tools, techniques, and methods of doing employed in task performance. Since each system operates according to its own kind of laws, each may be treated as an independent system.

Characteristics of the Technological System

The technological system, because of its inanimate character, is a reactive system. It is not able to act independently, but only in reaction to the behavior of the social system. Since people are required to initiate and control the activities of the technological system, it is customarily assumed that it is a passive system with relatively minor effects on the functioning of the social system. Beyond the purely technical requirements that must be met to operate a particular technology effectively—e.g., the speeds required to operate a machine at an optimum rate—there are a number of other characteristics that have important consequences for the social system. These characteristics emanate from the inanimate laws that govern the behavior of the technological system, and they have a major impact on the animate characteristics of the social system—i.e., its biological and psycho-social dimensions. These latter kinds of technological consequences emerge when we are concerned with "the problems of relating the technological requirements to men as self-willed beings, not simply as another kind of machine, and to groups of men, not simply to isolated individuals" (Emery, 1959, p. 10). Although the impact of the technological system varies according to the specific technology employed and the way it is organized for productive effort, the following characteristics appear to have significant consequences for social systems over a wide range of work settings: (1) the characteristics of the material being produced, (2) the physical work setting, (3) the spatio-temporal dimension, (4) the level of mechanization,

(5) the unit operations, and (6) the degree of centrality of different operations (Emery, 1959).

The Characteristics of the Material Being Produced

The natural characteristics of the material being produced may have a significant influence on the social system. This dimension is critical to the extent that it introduces uncontrolled variation into the labor requirements of the production process; for example, the kind of metal alloy that enters a series of interdependent machining operations may influence the strain experienced by machine operators. One form of alloy may lead to a relatively easy process of waste removal, while another may clog the machines, thereby making waste removal difficult. The nature of the material can also produce problems of variance control in service-oriented industries such as education and health care. For example, variations in the attributes and behavior of patients often introduce a number of unintended consequences for medical personnel.

The Physical Work Setting

Light, temperature, noise, pollution, and orderliness provide a physical setting for work. Although these physical conditions are often dependent upon the material worked on and the level of mechanization, there is considerable variation within these broad limits to suggest that the physical work setting influences the social system, independent of the other technological dimensions. Herzberg (1966) proposes that these 'hygiene' factors, though not direct sources of motivation, can produce conflict among workers. Emery (1959) describes studies where management's concern or indifference to physical conditions can lead to positive or negative consequences for workers (p. 11). He also cites research demonstrating the direct impact of these conditions on productivity, health, and accident rates. Although the overall effects are just starting to be studied, available research clearly suggests that physical conditions do have important consequences for the behavior and attitudes of individuals and work groups.

The Spatio-Temporal Dimension

The spatial layout and the spread of the production process over time have significant consequences on the social dynamics of work. Coordination, mutual support, and interpersonal contact may be either facilitated or hindered by the spatio-temporal distribution of machines and production processes. Conant and Kilbridge (1965) have shown the reduction in interpersonal relations that results when work is changed from an assembly-line method of production to a bench assembly method. Trist *et al.* (1963) has shown the problems of coordinating interdependent operations over different work shifts. Recent studies in social ecology have also reported the impact of spatial and temporal configurations on the attitudes and behavior of individuals and work groups (Sommer, 1969; Steele, 1973). The accumulated evidence strongly suggests that this dimension contributes towards a specific social ecology that may have important consequences on the social processes involved in operating a particular technology. Much like the other technological dimensions, the extent to which spatio-temporal configurations make a real difference in social processes depends on a variety of contributing factors. Placing independent fabricating operations closer together, for example, may not increase interpersonal communication if such interpersonal contact is not required for task accomplishment.

The Level of Mechanization

Referring to the relative contribution made by machines and men to the transformation of raw material, historically there has been an increase in the level of mechanization toward greater automation. The historical review of work outlined the social effects of this trend: Attention shifts from the individual worker to the machine, human contributions to the production process lessen, and individuals are relegated to quasi-mechanical operations dictated by the requirements of a mechanized technology. Although there is some evidence that increased automation may place a greater demand on the decision-making and information processing capabilities of workers, the degree to which this will enhance the human element of work may depend on parallel changes in work design and managerial philosophies (Blauner, 1965; Shepard, 1971).

The Unit Operations

The phases of production required to complete the changes involved in producing goods and services have an impact on the labor requirements of the social system. These natural task groupings present individuals with certain work patterns frequently requiring different methods of coordination and production control. A primary difference between assembly-line production and continuous process techniques involves the methods employed to control the production process. In the former instance, workers attempt to contain and control sources of production variance within discrete segments of the manufacturing process, while in the latter, individuals control one or more of the conditions—e.g., rate of flow, temperature—that enable the production process to carry through to completion. The design of unit operations is often associated with different configurations of the other technological dimensions, therefore one would expect different task groupings depending upon various combinations of the other technological characteristics—e.g., the nature of raw materials, physical conditions of work, spatio-temporal dimensions, and level of mechanization.

The Degree of Centrality of Different Operations

This technological characteristic refers to the degree to which different operations in the total production process demand special attention, effort, or skill. Those operations requiring special attention may be considered central or necessary to the successful completion of the whole task. In treating a hospitalized patient, for example, it may be only treatments administered by physicians that require special skill, rather than the multitude of housekeeping and patient care activities also performed. The degree of centrality of different operations is frequently reflected in the organization of worker and supervisory roles. It is often possible to infer which operations are perceived as central or optional to task completion by observing how individuals differentially relate to different parts of the production process. Those operations that require high degrees of attention or skill should be central, while the others may be considered optional to the total production process. These latter operations are not really necessary to effective performance, but they are usually carried out for some real or presumed purpose.

Summary

The technological system is composed of the tools, techniques, and methods of doing employed by human beings for productive achievement. These factors are governed by physical laws in contrast to the social system which operates according to laws related to the animate character of social beings. The technological system is also a reactive system in that people are required to initiate and control its activities. Beyond the purely technical requirements needed to operate a particular technology, there are a number of other characteristics that exert a significant influence on the social system. These characteristics—e.g., the physical work setting, spatio-temporal dimensions, the level of mechanization—present workers with certain technological facts or realities that may facilitate or hinder the social processes required to run the technological system. Whether or not characteristics of the technological system interfere with or help the functioning of the social system depends on the way the relationship between the two systems is structured. Since the production of goods and services depends on the joint operation of both social and technological systems, it is the relationship or interface between them that is critical to the performance of work. Therefore, the fundamental issue is *how to integrate the two independent but related social and technological systems into an overall framework for effective task accomplishment.*

RELATING THE SOCIAL AND TECHNOLOGICAL SYSTEMS

In discussing the social and technological systems as independent elements of work, it is apparent that in any purposive work relationship in which human beings are required to perform tasks, there is a joint system operating, *a socio-technical system* (Davis and Trist, 1972). Basically, a socio-technical system is some organized collection of men and technology structured to produce a specified outcome. The achievement of this outcome depends upon the appropriate joint operation of both the social and technological components. Specifically, "to make the system work, the problem is to get the things [the technology] whose behavior is governed by one set of laws, to interface effectively with the men [the social system], whose behavior is governed by another set of laws" (Vaill, 1973, p. 2). The problem of effectively relating the social system with the technological system is "neither that of simply 'adjusting' people to technology nor technology to people but organizing the interface so that the best match [can] be obtained between both" (Trist, 1970, p. 13).

Joint Optimization: Directive Correlation

The effective matching of the social and technological systems to produce a product or service may be considered in terms of a joint socio-technical system. When the operation of such a system depends upon the combined operation of its social and technological parts, we are concerned with how these parts function jointly to produce an overall outcome. Specifically, the optimization of the system taken as a whole—the socio-technical system—depends upon the "joint optimization" of its independent yet correlative social and technical parts. Each part must function optimally if the work system as a whole is to operate effectively.

The term joint optimization derives from Sommerhoff's (1969) theory of directive correlation. Directive correlation is a theoretical construct that explains purposeful behavior as the relationship between two independent but correlative systems: the system or organism that seeks a purpose or goal and its environment. For our purposes, directive correlation describes the conditions that must be satisfied if the independent but correlative social and technical systems are to interact purposefully to produce a desired outcome. Briefly stated, directive correlation analyzes purposeful behavior into three successive time frames, T_0, T_1, and T_2. The first time frame, T_0, exists prior to the matching of the two systems to produce a goal or purposeful outcome. At this point, the two systems, the social and technical systems, for example, interact with one another such that there is a movement of information between them; this information informs the goal-seeking system (the social system) of an appropriate course of action to take at some later time (T_1) to respond to events which have in the meantime occurred in the environment (the technical system) if a goal is to be attained. The second time frame (T_1) designates that point at which the goal-seeking system (the social system) uses this information to respond to or match its environment (the technical system) to produce a desired outcome. Finally, the third time frame (T_2) represents the outcome of this joint system-plus-environment match, the goal state or focal condition.

An example should clarify the concept of directive correlation as it applies to socio-technical systems. Figure 2 illustrates a directive correlation between a machine operator and a milling machine. For purposes of explanation, let us assume the following: (1) the objective of the coupled, man-machine system is to remove excess metal from a wheel-shaped disk, (2) the speed of the machine directly varies with the hardness of the metal, (3) the machine must operate at a certain speed to successfully mill the disk, and (4) the operator can adjust the speed of the machine upward or downward to maintain the correct speed. We can now state the properties necessary for the purposeful behavior of the man-machine, socio-technical system. In order to bring about a successful outcome—that is, to remove excess metal from the disk at T_2—the machine operator must increase or decrease the speed of the machine, depending on how fast the machine is operating in relation to the metal, at T_1. The only way the operator could effectively adjust the speed of the machine at T_1 would be if he received information as to speed of the machine at some time prior (T_0) to his adjustments. This

Figure 2

DIRECTIVE CORRELATION BETWEEN MILLING MACHINE AND A MACHINE OPERATOR

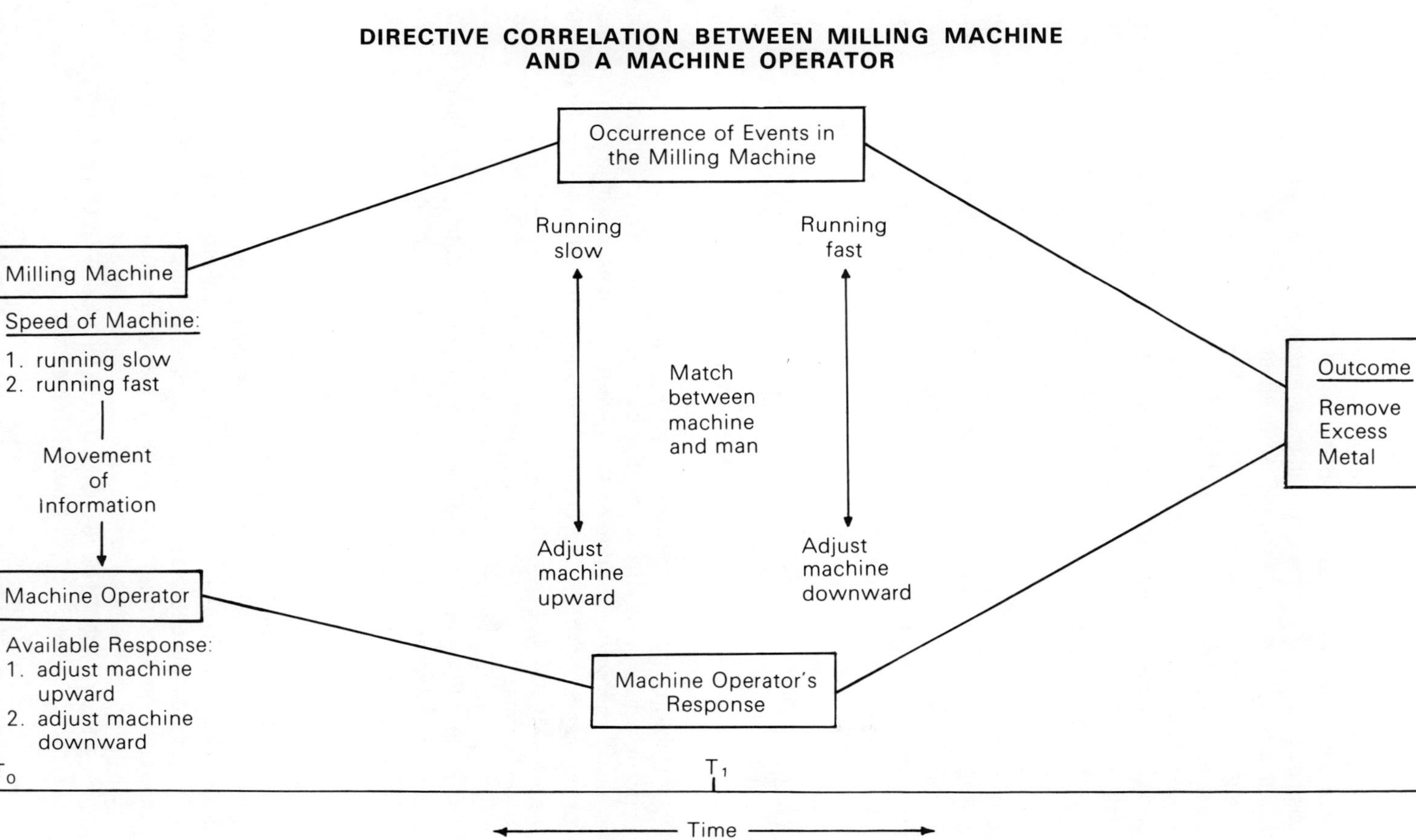

information at T_0 would have to inform the operator of the appropriate adjustment to make at some later time T_1 to respond to events (the speed of the machine) which had in the meantime occurred in the machine. This information would also have to be received long enough before the machine sped up or slowed down to enable the man to make the necessary adjustment. If excess metal is successfully removed at T_2 *and* if the above conditions are met, we can then say that there has been a successful match or directive correlation between the man and the machine.

Returning to the fundamental concept of a socio-technical system as composed of two independent yet correlative systems, directive correlation provides the theoretical framework for integrating both social and technological systems into a jointly optimized work system. Specifically, "where the achievement of an objective is dependent upon *independent* but *correlative* systems, then it is impossible to optimize for overall performance without seeking to *jointly optimize* these correlative systems" (Emery, 1969, p. 119). Since the social and technological systems are governed by different kinds of laws, joint optimization requires a match between the two systems such that each system functions optimally according to its own laws without interfering with the other system. In other terms, this match or directive correlation requires that the task requirements of the technological system and the biological and socio-psychological needs of the social system be jointly satisfied.

Summary

The idea that work is an agreement between two or more persons to perform a given task has led to the concept of a socio-technical system to represent the work structure existing when social beings relate to technology for purposeful results. The concept of socio-technical system implies that both social and technological systems relate to one another in some systematic manner to produce a specified outcome. The hyphen between the words socio and technical represents the directive correlation or matching that must occur between the independent but correlative systems to reach a desired outcome, while system signifies that this matching process forms a combined or coupled relation that itself can be considered an organized whole, a socio-technical system.

A jointly optimized relation between the social and the technological systems is a necessary but not sufficient condition for effective task performance. To perform a task effectively, a socio-technical system must also relate to relevant segments of its environment. Therefore, beyond the initial concern of relating the social system to the technological system so that a jointly optimized work system is formed, we are also concerned with relating the combined socio-technical system to its wider environment.

RELATING THE SOCIO-TECHNICAL SYSTEM TO ITS ENVIRONMENT

When we refer to the relation between a system and its environment, we make two basic assumptions; first that the system has been defined in a way that differentiates it from its environment and second that the system is relatively "open" with respect to its environment. Since it is not possible to relate two things to each other without first differentiating between them, socio-technical systems have to be defined as distinct from the environment. We must then determine if such systems are open or closed in respect to the environment. The choice of an open or closed-system perspective has a significant impact on how the system and environment relationship is conceptualized and how the behavior of the system is understood.

Definition of a System

A system is usually defined as "a set of objects together with the relationships between the objects and between their attributes" (Hall and Fagen, 1956, p. 18). This definition implies that a system is the sum of it's objects, relationships, and attributes; a system is composed of objects which possess certain attributes, and the objects along with their attributes are interrelated. While this definition accounts for the interdependence of the parts of a system, it fails to recognize that the objects or parts are not significantly connected with one another except in reference to a superordinate whole of which they are a member. To refer to parts, attributes, and relationships without referring them back to the whole or total system of which they are a part leads to the unanswerable question: Why are the parts related to one another? The members of a work team, for example, are related to each other only as they function to maintain a superordinate whole—a work team. Without the whole or work team as a reference point, the members, their attributes and their relationships have no organized meaning. Feibleman and Friend (1945) summarize this argument in relation to the structure of organization: "For purposes of analysis of structure no more is required than the whole from which the analysis starts and two levels of analysis [parts and subparts]" (p. 20). The whole is one level above its analytic parts and subparts. Therefore, reference to the whole is the starting point for understanding a system; parts, attributes, and relationships are redeployed at the analytic levels to describe the behavior of the system.

On the basis of the foregoing discussion the concept of system requires a logic of thinking that goes beyond relational thinking to encompass the treatment of wholes. A "holistic" framework starts from the premise that a system cannot be derived from its constituent parts; instead, the system is an independent framework in which the parts are organized. The system provides the framework or organizing principle for arranging the parts, and this arrangement produces a concrete organized object or whole (Angyal, 1941). Based on our earlier discussion that socio-technical systems are organized wholes, it is imperative that we define such systems in a way preserving their wholeness while differentiating them from the environment. It is also necessary to provide a definition describing the general class of all socio-technical systems while realizing that the identification of any concrete or real system requires a set of empirical criteria for discovering the boundaries of the system. (A discussion of criteria for bounding a concrete socio-technical system appears in the next chapter under the heading, Bounding a System.)

Miller (1965) provides a definition for a classification of concrete or real systems that meets the above requirements. He states: "A concrete, real, or veridical system is a nonrandom accumulation of matter-energy, in a region in physical space-time, which is nonrandomly organized into coacting, interrelated subsystems or components" (p. 202). Concrete systems are organized wholes existing in the real world. They can be differentiated from nonorganized entities in the environment by certain empirical operations carried out by outside observers. The physical proximity of units and the common fate of units are criteria that are frequently used to identify concrete systems. Since the classification of concrete system applies to a vast array of organized wholes existing in the real world, it is necessary to modify Miller's definition to delineate only those wholes that are socio-technical systems. This leads to the following definition: *A socio-technical system is a nonrandom distribution of social and technological components that coact in physical space-time for a specified purpose.*

This definition emphasizes the wholeness of socio-technical systems, because as nonrandom distributions of components in physical space-time, they exist as organized entities. It also underscores the notion that both social and technological components coact for a specified purpose, though the components may or may not be spatially contiguous during the interaction. Some socio-technical systems are located in relatively confined areas where all the components are physically together; other socio-technical systems are relatively dispersed over a wide geographical area. Obviously, modern information technology reduces the need to have coacting components physically together. Finally, the definition serves to differentiate the general class of all socio-technical systems from other organized and nonorganized entities in the environment. If a nonrandom distribution of components exists in a region in physical space-time, it is a whole; if this whole is composed of social and technological components which coact for a specified purpose, it is a socio-technical system. Based on this definition we can now address the issue of whether or not socio-technical systems are relatively open with respect to the environment.

Open vs. Closed Systems

The determination of whether a system is open or closed in respect to the environment depends upon the degree to which the boundary of the system is permeable to inputs of matter-energy and information from the environment. No concrete system is totally open or closed. A totally open system would not be selective in respect to inputs and, therefore, would not be independent from the environment, i.e., it could not be differentiated or bounded apart from the environment. On the other hand, a totally closed system would have impermeable boundaries and, therefore, would have to exist in a vacuum with no contact with the environment; i.e., it could not exist in the real world. Thus, the issue of whether a concrete system is open or closed is always relative.

The choice of treating a system as relatively open or closed depends on a number of assumptions about such systems. Briefly, open systems require inputs of matter-energy or information from the environment to maintain themselves in states of relatively high energy and complexity. Closed systems do not require such inputs, and they are not able to maintain such states. Open systems carry out exchanges with the environment by self-regulating their activities to achieve highly-organized, steady states that enable them to do work despite fluctuations in the environment. Closed systems do not achieve steady states but tend toward an equilibrium where the system is unorganized and unable to do work. Finally, to the extent open systems maintain effective relations with the environment, they are able to grow and to develop toward greater levels of complexity and heterogeneity of parts. Closed systems tend in the opposite direction.

These differences lead to a number of consequences for understanding the behavior of systems. The first concerns the point of reference for understanding system behavior. An open system perspective directs attention to system and environment interdependence; to understand the system it is necessary to understand the relevant environmental forces. A closed system approach seeks to understand system behavior within the context of the system; behavior is explained in terms of the internal structure of the system, since the environment is viewed as relatively unimportant for system functions.

The second consequence of open or closed system assumptions involves the notion of variance or irregularity within the system. An open system perspective views variance as essential to the self-regulating properties of the system. Open systems require certain amounts of variability in their behavior in order to adjust to environmental changes. Growth towards greater complexity and heterogeneity is seen as part of the natural development and survival of the system as it seeks to maintain effective environmental exchanges. A closed system perspective views variance within the system as disruptive to system functioning; therefore, it should be controlled or minimized to enable the system to remain in a stable state.

The third consequence is concerned with the amount of certainty experienced by the system. Since open systems are subject to environmental influences, they are always faced with some degree of uncertainty as to what behavior is most appropriate. Instead of maximizing their choices, open systems "satisfice" by doing the best that they can in the face of uncertainty (March and Simon, 1958). Based on this assumption, an open system perspective stresses the importance of obtaining adequate information from the environment. It also emphasizes the significance of the system actively influencing the environment as a means of bringing about a more certain state of affairs. A closed system approach focuses on the certainty inherent in systems that are relatively unaffected by the environment. Given the information that exists within the system, a closed system perspective assumes a high degree of rationality in respect to system behavior.

The final consequence that follows from open or closed system assumptions involves the structure of systems. In relating to a changing environment, open systems are capable of producing a given outcome in a variety of ways—that is, they can modify their structures to reach the same objective. Thus, there is no one-best-way to structure the system to produce an outcome; instead, the effectiveness of a particular structure always depends on the particular environmental circumstances encountered by the system. A closed system perspective, on the other hand, directs attention to the one-best-way of structuring the system. Since this approach assumes a high degree of certainty and rationality within the system as well as minimal amounts of influence from the environment, it seeks a maximally structured system to produce a specified outcome.

The question of choosing an open or closed system perspective poses an interesting problem. If we choose one approach rather than the other, our understanding may be limited to only a part of the system's behavior. Thus an open system perspective focuses on the uncertainty involved in system and environment relationships, while a closed system approach draws attention to the certainty required for rational system functioning. Though both approaches are valid, each deals with only a part of the total system. The answer is not to choose one perspective over the other but to incorporate both into an understanding of socio-technical systems.

Thompson (1967) arrives at a similar conclusion in relation to complex organizations: "We will conceive of complex organizations as open systems, hence indeterminate and faced with uncertainty, but at the same time as subject to criteria of rationality and hence needing determinaters and certainty" (p. 10). This definition emphasizes the opposing forces of certainty and uncertainty that coexist in organizations. It also emphasizes that both open and closed system perspectives are required to understand such systems. Therefore, we will treat socio-technical systems as open systems, subject to environmental uncertainty, but at the same time as requiring certainty to function rationally.

Summary

Two basic assumptions underlie the issue of relating a system to its environment. The first is defining the system in a way that differentiates it from the environment. Based on Miller's (1965) definition of a concrete system, a socio-technical system is a nonrandom distribution of social and technological components coexisting in physical space-time for a specified purpose. This definition emphasizes the wholeness of such systems and distinguishes them from their environment. The second assumption is whether the system is open or closed with respect to the environment. Since socio-technical systems display both open and closed system behavior, it is necessary to use both perspectives to understand such systems. Thus, socio-technical systems are conceived not only as open systems—subject to environmental uncertainty—but also as requiring certainty to function rationally.

The above assumptions provide the groundwork for relating socio-technical systems to their environments. When taken together, the assumptions appear to be contradictory: as closed systems socio-technical systems are relatively independent from the environment, but as open systems they are relatively interdependent with it. To reconcile this paradox, it is necessary to understand how such systems maintain themselves in organized states while interacting with their environment. This requires knowledge of the properties of open systems and how it applies to socio-technical systems.

Properties of Open Systems

Open systems are concrete wholes that interact with their environment. Their existence depends upon a continuous exchange of energy with relevant segments of the environment. The environment provides inputs of energy; the system uses these inputs to maintain itself and to produce outputs of energy; and the environment absorbs these outputs and provides new inputs. This energy reinforcing cycle continues as long as the environment provides the system with required inputs and the system furnishes the environment with needed outputs. The modern business organization typifies this interdependence. The organization purchases raw materials from vendors in the environment, it converts the raw materials into finished goods and markets them in certain sectors of the environment, and these goods provide the environment with needed products and furnish the organization with a means of obtaining additional raw materials. Since these energy exchanges enable both the system and the environment to continually replenish themselves, open systems have no existence apart from their environment, and the environment has no relevance apart from the systems that interact with it. Therefore, when we speak of an open system, we must specify its environment, and when we speak of an environment, we must specify the system(s) it environs or surrounds.

Open systems also display a hierarchical ordering, each higher level of system being composed of systems of lower levels. Systems at the level of society are composed of organizations; organizations, of groups; groups, of individuals, and so on. Though it is possible to identify several levels of systems, we are concerned with socio-technical systems at three levels:

Table 1

HIERARCHICAL ORDERING OF SOCIO-TECHNICAL SYSTEMS

System Level	Composition of Parts	Example
Individual	One individual and the technology he uses for production	Man-machine system
Group	Several interrelated individuals and the technology they use for production	Work team
Organization	Several interrelated groups and the technology they use for production	Business organization

individual, group, and organization. Table 1 illustrates this hierarchical ordering. Systems at each level consist of both social and technological components organized for a specified purpose. A simple man-machine system typifies the lowest level, while a work team and a business organization illustrate the other levels. The hierarchical nature of open systems emphasizes their joint independent and interdependent nature. An open system is both an independent framework for organizing lower level systems and an interdependent member of a higher level system. Thus, a system's autonomy in organizing its parts is always tempered by the controlling forces of the higher level system of which it is a member.

Though systems at various levels differ in many ways—size and complexity for example—they have a number of common characteristics by virtue of being open systems. Since these characteristics apply to open systems at all levels, they may be considered cross-level explanations of how such systems function. The following five properties explain how open systems maintain themselves in relatively independent states while interacting with their environment: (1) import-conversion-export cycle, (2) boundary, (3) steady state, (4) regulation and control, and (5) equifinality.

Import-Conversion-Export Cycle

An open system exists by exchanging forms of energy with its environment. Figure 3 portrays this energy exchange as a four-step cycle in which

Figure 3

THE FOUR STEPS OF THE IMPORT-CONVERSION-EXPORT CYCLE

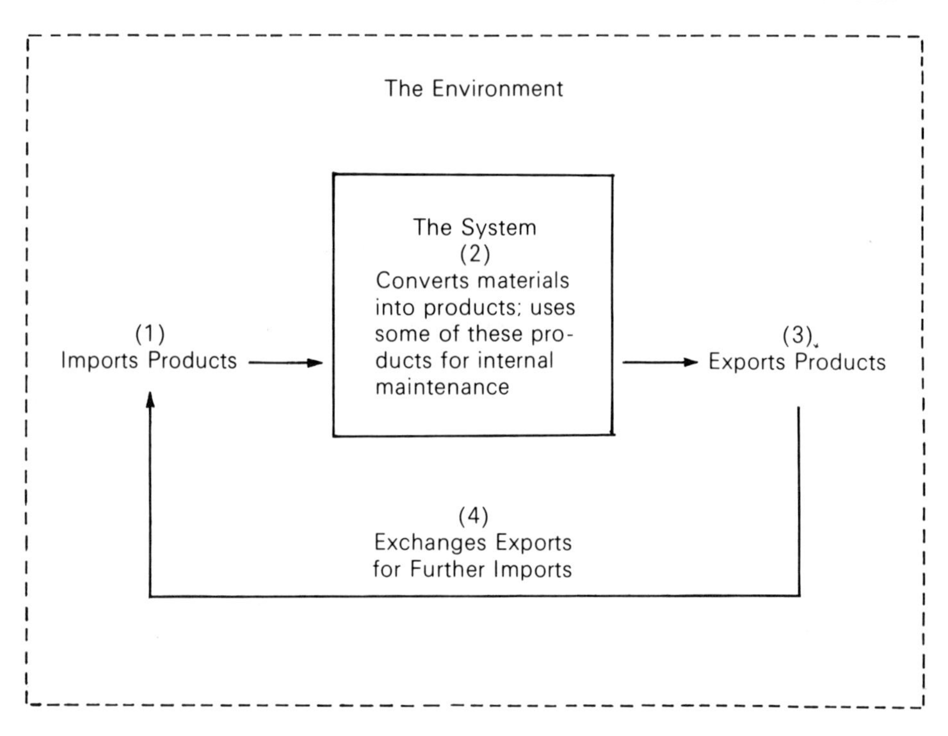

the system: (1) *imports* materials from its environment, (2) *converts* them into products and in this process consumes some of the products for its internal maintenance, (3) *exports* the rest of the products back to its environment, and (4) directly or indirectly, *exchanges* its outputs for further inputs.

The import-conversion-export cycle may be considered the work a system must perform if it is to exist. Often, when open systems appear to be at complete rest—a person sleeping for example—they are doing work to maintain themselves. A system is likely to persist and grow as long as the energy expended in doing work is less than the amount obtained from the environment. If this ratio is reversed for any length of time, a system gradually exhausts its energy reserves and ceases to exist.

Socio-technical systems, like all open systems, require a continual supply of energy. They import a variety of energy sources from their environments—raw materials, utilities, and supplies. They convert these by means of a joint social and technological conversion process into products or services, and they export some of these back to their environment. In most cases, money is used to represent energy units, and the flow of money—from buying to producing to selling—symbolizes the energy cycle.

In charting the energy cycle of most socio-technical systems, there are several import-conversion-export processes operating. Generally, as socio-technical systems become more complex—from simple man-machine systems to organization-level industries—the number of separate or related energy cycles increases. Thus most hospitals process food and linen in addition to patients. Each of these conversion processes is associated with a particular set of inputs and outputs, and in most cases, a specific set of social and technological components. Although each energy cycle may contribute to the objectives of the system, it is often difficult to understand what actual process is essential for the system's survival; without this knowledge, we are faced with the problem of determining the basis for the system's existence. In the case of a hospital, the system could be a food processing, dry cleaning, or medical care organization depending upon the particular energy cycle chosen. To overcome this problem we use the concept *primary task* to define the conversion process that the system "must perform if it is to survive" (Miller and Rice, 1967, p. 25).

The primary task is a heuristic concept allowing us to sort out from among the variety of activities taking place in a socio-technical system the critical processes that the system must carry out in order to exist. The definition of the primary task determines the system's dominant import-conversion-export cycle as well as the social and technological components essential for its survival. Since the primary task may change over time, it is necessary to postulate that at any given time a socio-technical system has only one primary task. Once the primary task is identified, all other tasks and the energy cycles associated with them, may be considered ancillary. If we can assume that the primary task for a hospital is providing medical services, we can then identify the dominant imports and exports related to this task and the social and technological resources required for its performance. Medical supplies, persons in need of medical services, new medical personnel, and innovative or replacement medical technology are examples of imports, while used or worn out medical supplies and technology, persons in relatively good health, and retired medical personnel represent some of the exports. Nurses, physicians, laboratory technicians, and other medical staff are the necessary social components, while operating tables, x-ray machines, traction devices, and other health care technology illustrate the appropriate technical resources. Those additional import-conversion-export cycles associated with food, linen, housekeeping, and record keeping represent ancillary tasks supporting the primary task and helping in its completion.

The concept of primary task is important for understanding socio-technical systems. It allows us to order the multiple activities taking place and to determine the priorities of constituent parts of the system. Defining the primary task frequently leads to conflicts within the system or between the system and its environment. In the former instance, a constituent part may define its primary task in a way which contradicts that of the system. A university faculty may define its task as generating new knowledge, while the administrative leaders might envision a greater pay-off in limiting this activity and increasing the educational opportunities for undergraduate students. Similarly, conflicts between the system and its environment may arise when each defines the system's primary task differently. A community, for example, may define the primary task of a city university as increasing the educational level of the local population. This definition may contradict the objective of the university to become a prestigious center for research and graduate study. In both examples, the definition of the primary task has a significant impact on the internal resources and environmental relationships of the system. Without an adequate task definition, socio-technical systems are unable to organize effectively their social and technological resources or their relationships with their environments.

Miller and Rice (1967) summarize the importance of an adequate task definition for socio-technical systems at the organizational level:

> The primary task is not a normative concept. We do not say that every enterprise must have a primary task or even that it must define its primary task; we put forward the proposition that every enterprise, or part of it, has, at any given moment, one task which is primary. What we also say, however, is that, if, through inadequate appraisal of internal resources and external forces, the leaders of an enterprise define the primary task in an inappropriate way, or the members—leaders and followers alike—do not agree on their definition, then the survival of the enterprise will be jeopardized. Moreover, if organization is regarded primarily as an instrument for task performance, we can add that, without adequate task definition, disorganization must occur (pp. 27–28).

The concept of primary task is helpful for understanding the relationship of a socio-technical system with its environment. Since the primary task enables us to identify the dominate energy cycle of the system, it delineates those points at which the system must relate to its environment if it is to survive. A hospital, for example, must relate to local physicians for patient referrals, to a pharmaceutical system for medical supplies, and to an available pool of medical personnel for qualified staff if it is to perform its primary task of providing medical services effectively. The conditions for interacting successfully with the environment lie both inside and outside the system. On the one side, the system must have at its disposal and be able to organize the necessary internal resources—both social and technological—for performance of the primary task. On the other side, there are independent environmental changes that may affect the system and render its internal resources ineffective. The ability of a socio-technical system to maintain itself in the face of these environment forces is, in large part, dependent upon its flexibility. Variations on the import and export sides of the system can be tolerated only to the extent that the social and technical resources are able to adjust to external influences. Thus, depending upon their flexibility, different combinations of inputs may be processed to yield similar outputs and different mixtures of outputs may be produced from similar inputs.

The flexibility of a socio-technical system may be limited by either its human or physical component. A machine that is designed to operate within a narrow range of speeds or a worker who is relatively unskilled or inexperienced may place a severe constraint on the whole system. The chief source of internal constraint for a socio-technical system depends upon its primary task. In those systems where the primary task depends heavily on the technical component—a highly automated oil refinery for example—the flexibility of the technology is the major source of constraint; in others—such as educational institutions or the psychiatric department of hospitals—people are key to primary task performance and their ability to adjust to external influences is the primary constraint. Yet in others, such as mechanized production lines or transportation systems—both people and machines are heavily involved in task performance and their joint flexibility determines how well the system accommodates to the environment.

The social and technological components, in performing the primary task, function as one of the major boundary conditions of the whole system. They help to determine what the inputs and outputs are. In carrying out the work of the system, the human and physical components set constraints on what the system can do and on what environmental forces the system can deal with effectively. Thus, in mediating between the objectives of the system and the environment, the social and technological resources help to define those conditions under which the system can maintain some degree of independence from its environment.

Boundary

The term boundary is commonly used in two ways. When applied to concrete systems that exist in regions in physical space-time, it has a quasi-geographical connotation; the boundary arises in physical and temporal space and serves to differentiate concrete objects that exist in physical spacetime. The geographical boundaries that separate one city, country, or continent from another during a specified period of time illustrate this kind of boundary. A second use of the term is when individuals abstract an attribute of a concrete system and use this abstraction to differentiate systems. Race, height, intelligence, sex, and education are examples of attributes often used to distinguish between one group and another. Both connotations of the concept boundary are useful for distinguishing between a system and its environment. The geographical usage enables us to spatially and temporally orient systems, while the abstract usage allows us to differentiate systems along almost any criterion which serves our purposes. To avoid confusion, we refer to boundaries that exist in physical space-time as "concrete boundaries" and to those that are based on abstracted criteria as "conceptual boundaries."

The concrete boundary of an open system is that region in physical space-time differentiating the system and its environment. A person's skin and clothing, a family's house, and a nation's frontier are examples of concrete boundaries. Since open systems exchange materials with their environments, the boundary is selective yet permeable. It filters out a variety of environmental forces while permitting a selective set of materials to enter or leave the system. In performing this dual function, the boundary maintains the integrity or independence of the system while allowing it to relate to its environment. When we say that a system has form or substance we imply that its boundary is distinct enough to permit us to distinguish between what is within and without the system. Because our sensory organs are attuned to placing objects into the concrete boundary of space and time, concrete boundaries are particularly helpful in distinguishing systems from their environment by orienting them accurately in relation to one another.

In contrast to a concrete boundary, the conceptual boundary of an open system exists only when someone decides on some criterion or attribute to differentiate systems. Using sex as a criterion, someone might decide that men and women are different, while somebody else using a different criterion–age, for instance—may not separate them. Since conceptual boundaries are based on abstractions, they need not be limited to physical and temporal space. Other kinds of "space" frequently used to differentiate systems include: social space—those attributes commonly possessed by all human beings such as weight, height, sex, ethnic background, race, and living area—and psychological space—those attributes of personality or behavior that lead people to perceive themselves as different from others, e.g., needs, values, and levels of psychological growth. Conceptual boundaries are useful because they permit us to delineate systems in ways that uniquely serve our purposes. If we are interested only in certain attributes of open systems—height and weight, for example—we can bound systems using these criteria without having to take into account their actual locations in physical space-time or the rest of the attributes they possess. Although this allows for a certain mental convenience, conceptual boundaries are often confusing and unnecessarily complicated. Thus, we employ conceptual boundaries with extreme caution, realizing that they are nothing more than mental constructs existing for a particular purpose.

In addition to the kind of boundary—concrete or conceptual—that one chooses to differentiate a system, open systems may also have external and internal boundaries. Systems containing two or more lower level systems—a group which contains at least two people, for example—have both external and internal boundaries. These boundaries may be either concrete or conceptual. If we are referring to concrete boundaries, the external boundary holds together the lower level systems which make up the system, protects them from environmental forces, and admits or excludes certain materials and information. The internal boundary provides these same functions for the lower level systems that comprise the larger system—it holds together their components, protects them from the environment, and selectively filters materials and information. The external boundary of a work group, for example, is the walls, ceiling, and floor of the group's geographical work area during a particular time span—e.g., 8:00 a.m. to 4:30 p.m. This spatial and temporal boundary holds the members together so they can directly interact, protects them from environmental influences such as excessive noise, rain, and temperature fluctuations, and admits or excludes certain materials and information. The group's internal boundary is the skin and clothing of its individual members. It serves the same functions as the external boundary, but its frame of reference is the individual members and their internal components. Similarly, if we are referring to conceptual boundaries, the external boundary differentiates the system and its environment, and the internal boundary distinguishes the system's components from each other. A work group's external boundary might, therefore, be defined by the attribute of skill level. Individuals above a certain skill level would be considered members of the group, while others would not. The group members could be further distinguished by defining their internal boundary on the basis of seniority. Thus, the members of the high skill-level group might be further separated into subgroups of low, medium, and high seniority.

Given the discussion of the kinds of boundaries that can be applied to open systems, let us relate the concept of boundary to socio-technical systems. The boundary of a socio-technical system distinguishes an interrelated set of social and technical components from its environment. Since such systems are concrete entities whose components coact in physical space-time, they always have a concrete boundary. Miller (1959) has proposed that the concrete boundary of a socio-technical system constitutes a differentiation of technology, territory, or time. This differentiation is necessary to distinguish the components and activities comprising the system from other components and activities making up the environment. Returning to the example of a hospital, the people and technology that provide medical services for infants in McDonald House on the campus of Case Western Reserve University at 9.00 a.m. on November 14, 1974, constitute a differentiated socio-technical system known as Rainbow Children's Hospital. This system may also have a variety of conceptual boundaries depending upon the purposes of those who enact such boundaries. Using age as a boundary criterion, for example, the people who work in this hospital can easily be distinguished from the patients.

Since socio-technical systems are composed of lower level social and technological systems, they possess both an external and internal boundary. The external boundary differentiates the system from its environment, and the internal boundary distinguishes the social and technological components from each other. In terms of concrete boundaries, these boundaries hold the parts of the system or components together, protect them from environmental disturbances, and selectively relate the system or its components to their respective environment. In most socio-technical systems these boundary functions are performed by physical artifacts such as buildings or fences and people such as inspectors or security guards. If the external and internal boundaries are conceptual, they serve as abstract definitions of the system and its components.

A primary function of the concrete boundary of a socio-technical system is to regulate exchanges with the environment. This regulatory function is related to the import-conversion-export cycle of the system. The boundary regulates the materials and information that enter and leave the system by serving as a region of control between the conversion process of the system and the environment. Since we have identified the primary task of a socio-technical system as that conversion process that the system must perform if it is to survive, the boundary can be considered the region of control regulating the imports and exports of the primary task. Figure 4 illustrates the relation of the boundary control function to the primary task.

To regulate environmental exchanges for the effective performance of the primary task, the boundary must selectively filter materials and information that enter and leave the system. This filtering process is geared to the specific requirements and capabilities of the social and technical components comprising the primary task—i.e., the social and technical resources that perform the primary task determine the boundary conditions or environmental exchanges that the boundary regulates. If these resources are relatively flexible in relation to imports and exports, the boundary can allow a wider range of environmental conditions to affect the system than if these resources are inflexible. A hospital, for instance, may be able to treat a wide variety of diseases with its current social and technological resources; thus, those who control the admission of new patients—a boundary control function—are

Figure 4

THE BOUNDARY CONTROL FUNCTION OF THE PRIMARY TASK

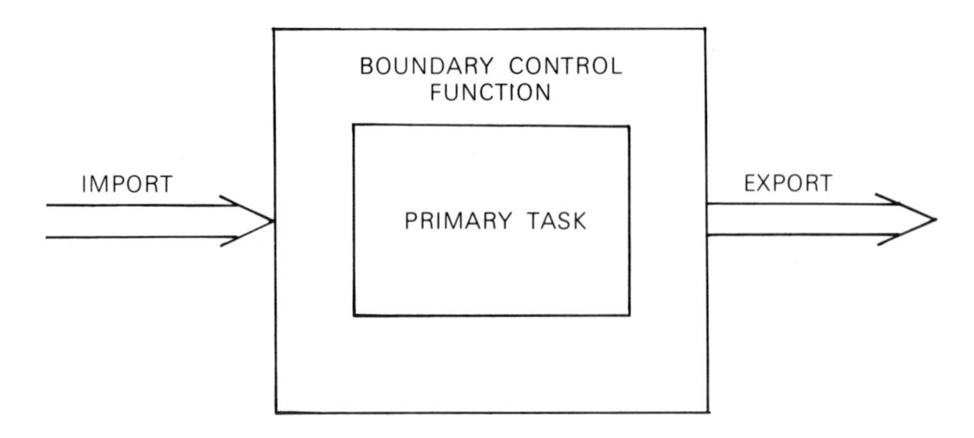

able to admit a wide range of imports to the primary task. Since the social and technical components serve as one of the major boundary conditions of the system, it is not relevant to define the boundary of a socio-technical system apart from the resources that carry out the primary task.

In summary, the boundary of a socio-technical system implies both a discontinuity of the primary task and an "interpolation of a region of control" (Miller and Rice, 1967). The discontinuity may take the form of a differentiation in technology, territory, and time, or in one of the other kinds of social or psychological space often used to differentiate one system from another. The region of control regulates exchanges with the environment; this regulatory function is guided by the requirements and capabilities of the social and technical resources performing the primary task of the system.

Steady State

Open systems are able to maintain their order and complexity while carrying out their energy cycles. This dynamic form of equilibrium is referred to as a steady state, and in contrast to the physical connotation of the term equilibrium in which a system is at complete rest, steady state denotes a system capable of preserving its character or integrity while doing work and relating to its environment. The activities involved in maintaining a steady state can be thought of as a continuous series of adjustment processes whereby the system keeps an orderly balance among its components and between itself and its environment. (The maintenance of a steady state is more fully discussed in the next section on regulation and control.) Forces that tend to disrupt a system's steady state—whether they emanate from within or without the system—are countered by the system's adjustment processes which restore as closely as possible the steady state. Since this adjustment activity enables a system to preserve its basic identity in the face of various disruptions, open systems are able to remain relatively stable while importing materials from the environment, converting them into products, and exporting the products back to the environment.

The existence of a steady state implies that exchange and conversion processes are operating within the limits considered adequate for the system's survival. In other words, certain critical variables having to do with environmental relationships and internal system functioning are being maintained within the ranges of stability required for the system's survival and growth. These steady state variables represent those conditions that the system must maintain if it is to exist. In human beings, for example, variables such as the rate of oxygen diffusion in the lungs, body temperature, blood acidity, and basil metabolism rate represent steady state variables that must be maintained within specific limits if life is to be sustained. Since each system has a unique set of steady state variables, they may be used to distinguish one system from another. Thus, a human being may be differentiated from a frog by virtue of the differences in the steady state variables and the ranges of stability that each maintains. Based on specific steady state variables and their ranges of stability, open systems use their adjustment processes to survive. When disruptive forces push a steady state variable beyond its range of stability, the system adjusts itself to eliminate these disturbances. If it is successful, the system maintains its basic character; if it is unsuccessful, it operates at a diminished state or terminates altogether.

Socio-technical systems exhibit steady state behavior when they are able to exchange with their environments and do work in accordance with their requirements for survival and growth. Like all open systems, socio-technical systems seek to maintain two related sets of steady state variables within the limits of stability required: those related to exchange processes—*import and export variables*; and those related to the internal functioning of the system—*conversion variables.* The import and export variables specify those conditions necessary for effective environmental relationships. The rates, frequency, cost, and kinds of imports and exports that enter and leave the system represent such variables. A hospital, for instance, has specific limits on the rate of new patients that can effectively be treated; a rate of new patients above or below this specified limit would render the system either over or underutilized, thus affecting its capacity for future environmental relations. Similarly, the conversion variables specify those conditions necessary for the effective transformation of imports into exports.

Since socio-technical conversion processes are composed of independent but correlative social and technological systems, the steady state variables associated with each of these resources are themselves, independent but correlative. They are independent by virtue of the fact that the behavior of both systems follows different kinds of laws. Thus, variables related to the steady state of the social system—rates, kinds, and intensity of social interaction for example—derive from laws of the animate world: such as biological and social/psychological laws; while variables associated with the steady state of the technical system—rate and frequency of producing for instance—derive from laws of the inanimate world, such as physical, electrical, and hydraulic laws. Both kinds of steady state variables are correlative in that both components must relate to one another to produce a desired outcome. Thus, to maintain its steady state, a socio-technical system must maintain the steady state conditions of both its social and technological components, such that each operates in accordance with its respective laws. In doing this, it must also assure that these steady conditions are maintained when the two systems are operating jointly to produce a specified output.

Let us return to the hospital example to describe more fully the interdependent nature of the steady states of the social and technological components. We will focus on a well-trained surgical team that is performing open-heart surgery. Briefly stated, the social system consists of two surgeons, three nurses, and an anesthesiologist, each performing their respective organizational roles. The technical system comprises a myriad of surgical tools and related medical supplies, a device to administer anesthesia, and a heart-lung machine to replace the function of the patient's heart during the operation.

When the entire socio-technical system is functioning properly, two related sets of steady state variables are operating within their respective limits: socially, the members of the team are performing their interrelated tasks at a rate and at a level of interaction necessary to complete the operation; technically, the surgical tools and related machinery are operating within certain prescribed limits to fulfill their functions. Each component is maintaining its steady state, and the combined system, when the components are performing jointly, is functioning within limits necessary to assure a successful outcome.

Now let us assume that one of the technical components—the heart-lung machine, for instance—malfunctions, thus throwing the steady state condition of the technical system into disorder. Immediately, specific members of the surgical team rush to disconnect the malfunctioning machine and to activate a back-up machine. During this time period, the steady state condition of the social system, expressed by the rate and level of task interaction, is also disrupted. Once the back-up machine is connected and the necessary adjustments are made, both social and technical systems gradually return to their previous steady states. This rather dramatic example demonstrates clearly the interdependent nature of the steady states of the social and technical components of a socio-technical system. Changes in the steady state condition of one component frequently necessitates modifications in the other if the system as a whole is to function properly.

To summarize our argument, open systems seek to maintain a complex set of interrelated variables in a steady state. These variables—related to exchange and conversion processes—represent those conditions that the system must keep relatively stable if it is to survive and to develop. Since the steady state of each system is relatively unique to its own internal make-up and its environmental situation, it is possible to distinguish one kind of system from another by virtue of its steady state variables. For socio-technical systems, the steady state variables derive from two independent but correlative social and technological components. If such systems are to function properly, they must maintain the steady state conditions of each component such that the state of the overall system is relatively stable when the components are working jointly toward a common outcome.

As already mentioned, open systems maintain their steady states through the use of adjustment processes whereby disruptions of steady state variables are counteracted. This adjustment activity can be considered a regulatory process whereby the system controls for deviations from a preferred steady state.

Regulation and Control

Open systems maintain their steady states through a process of regulation and control. They adjust their behavior to keep steady state variables within ranges of stability considered adequate for the system's survival. Regulation involves a feedback process whereby information about the current condition of steady state variables is compared against a standard of stability. If there is little or no deviation from such standards, the system is functioning properly, but if a steady state variable is functioning beyond its range of stability, the system must institute an adjustment process to correct for this discrepancy. The use of feedback to maintain a steady state is referred to as *negative feedback*, since the system uses this information to dampen or negate deviations from normal ranges of stability. In human beings, for instance, a negative feedback process is the control of body temperature somewhere close to the value of 98.6 (F). Increases or decreases in body temperature about this point are counteracted by various adjustment processes—e.g., perspiration, rate of heart beat, muscle contractions. The opposite process of *positive feedback* refers to those instances where the system employs feedback to increase deviations from a steady state. A positive feedback process would be the spread of a cancerous growth. The success of initial cancerous cells amplifies the growth of subsequent cells such that the entire growth increases at an increasing rate. Although positive feedback frequently has deliterious consequences for a system, some amount of this kind of feedback is required if systems are to alter their current steady states and move to higher-level conditions. The critical problem is to employ positive feedback in a way that initiates system changes effectively without altering the value of steady state variables to such an extent that the system is destroyed.

Since the steady states of all open systems are regulated by negative feedback, a discussion of the requirements necessary for successful regulation should provide a better understanding of this kind of feedback process. Cybernetics, the study of feedback control, has identified the following characteristics as necessary for successful regulation:

1. A set of steady state variables and their ranges of stability which serve as standards for system performance.
2. A method for obtaining information as to the actual state of these steady state variables so that comparisons can be made against those standards.
3. A repertoire of behaviors that can be employed to correct for any deviations from the standards that are found.
4. A method of decision-making and acting that enables the system to choose and enact a corrective course of action before the cause of the disturbance changes.

Each of the above characteristics implies certain assumptions about the regulation and control of open systems. First, a set of steady state variables that acts as a standard for system performance is a necessary requirement for goal-directed behavior. The steady state variables can be considered goals, and "it is the deviations from [these goals] that direct the behavior of the system, rather than some predetermined internal mechanism that acts blindly" (Buckley, 1967, p. 54). When viewed from this perspective, a system's goals represent a preferred steady state that the system acts to attain. Deviations from these goals require various adjustment behaviors as the system attempts to reduce any discrepancy between its present state and the preferred goal state. Since the actual attainment of a preferred goal state—where there is no discrepancy between present and preferred goal states—would imply that the system has no need for adjustment or goal-directed behavior, a system's goals are never totally attained if the system is to continue to act in a goal-directed manner. Thus, the goals of a system, in the form of preferred steady states, serve as intended future states of the system, and all adjustment processes to reduce deviations from these states represent goal-directed behavior.

The second implication involves the information required if the system is to learn about its deviations from its preferred steady state. Knowledge of how well a system is doing in relation to a standard of performance requires a method for monitoring the outputs or behavior of the system and using this information to adjust subsequent inputs. Commonly referred to as a negative feedback cycle or loop, this four-step process is shown in Figure 5 as follows: (1) information about the output or behavior of the system is fed back to a comparison device, (2) the comparison device compares this information against the preferred steady state and deviations signal a need for adjustment activity, (3) a decision-making device uses this information to synthesize a new set of behaviors to reduce any discrepancy, and (4) this new information is fed into the system to inform the system how to change its behavior in accordance with the decision-maker's instructions. This process continues until a steady state condition is reached for a particular steady state variable. The feedback loop may pass inside or outside of the system, the former constituting internal feedback and the latter representing external feedback from the environment. Regardless of the source of information, all feedback has a certain "probability of error," a specific "lag" in the time that it takes for a system to affect corrective behavior, and a certain "gain" which is the extent of corrective effect. Thus, although information is the essence of the feedback process, depending upon its probability of error, its lag time, and its extent of gain, there is a great deal of variance in how well systems can use it to maintain their steady states.

A third implication of the characteristics for effective regulation concerns the behavior that is needed to correct for deviations from a steady state. Once a disturbance disrupts a steady state, the system must be able to mount a counter-attack, as it were, to nullify the consequences of this disturbance. This corrective behavior must match the specific disturbance, in that it must be able to deal with the underlying causes of the disturbance. If the system cannot affect this match, it runs the risk of being destroyed by uncontrolled forces. Ashby (1966) develops this notion into a general requirement for effective regulation: The Law of Requisite Variety. Briefly, Ashby's law states that systems must process a requisite variety of corrective responses to match the variety of disturbances that affect the system. In short, "only variety can destroy variety" (p. 207). This requisite number of responses is not, however, a sufficient condition for effective control. The system must also contain responses that are effective for controlling the disturbances encountered. Thus, the system must not only possess a requisite number of responses, but among these must be an effective response.

Ashby's law is an elegant and useful concept for understanding how a system might increase its regulating ability. Internally, a system can increase its own variety of corrective responses. This would make the system more adaptable to a wider range of potential disturbances. Increasing the variety of corrective responses can be considered a learning process whereby

Figure 5

NEGATIVE FEEDBACK CYCLE

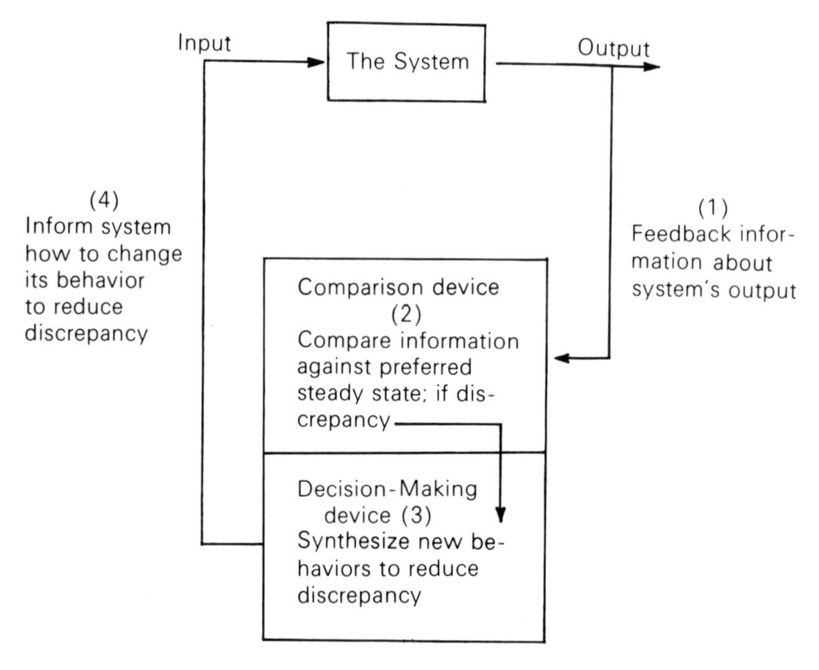

the system acquires new behavior to cope with emergent environmental forces. A second way for a system to increase its regulatory ability concerns the environment. Externally, a system can attempt to reduce the variety of disturbances in the environment. This would make the system susceptible to a narrower range of environmental forces. To simplify the environment, the system can either ignore certain external factors, or it can actively attempt to reduce the full range of actual or potential disturbances. The former strategy is tantamount to "sticking one's head in the sand," and though it may be useful for short-term regulation, the longer-term consequences of ignoring external forces would most certainly reduce the system's probability of survival. The latter approach, of actively reducing the environment's range of disturbances, appears to offer the system a better strategy for reducing the impact of environmental forces. Ideally, the system would attempt to intervene in the external dynamics of the environment so that sources of disruption would be nullified or simplified before they reach the system. Various forms of advertising and political lobbying are examples of organization-level systems' attempts to reduce external influences. In sum, Ashby's Law of Requisite Variety points to both internal and external opportunities for increasing a system's regulatory capacity.

The final implication of these four characteristics for successful regulation involves the method by which systems choose and implement various corrective behaviors. Regardless of the size or complexity of the system, regulation requires an effective decision-making and implementation process. The system, when informed of a deviation from a preferred steady state, must be able to analyze the cause of the disturbance, choose an appropriate corrective behavior, and implement this behavior before the cause of the initial disturbance changes. If the system fails to carry out this process at a rate at least equaling the change rate of the disturbance, the system will pursue a response that is no longer adequate for the circumstances at hand. Thus, to be systematically late may also be systematically wrong.

The effectiveness of a system's decision-making and implementation capacity depends, in part, on the state of the system's environment. Let us examine the environment in terms of its rate of change and the degree that its parts are connected to understand how the environment influences a system's regulatory process. Environments with slow rates of change and low degrees of interconnections of parts are rather simple in terms of system regulation. A slow change rate allows the system to predict the cause of a disturbance far enough in advance to allow the system to implement an appropriate course of action; while a low degree of interconnections permit the system to detect or separate the cause of the disturbance from the complexity of dynamics operating in the environment, since the causal networks existing under such circumstances are confined to isolated parts of the environment. Under these conditions, systems can anticipate and react to a predictable set of forces which emerge from the environment.

Complex environments, however, present a different set of circumstances. In contrast to more simple environments, complex environments have a fast rate of change and a high degree of interconnections. These conditions present systems with two major problems. First, predictions about the cause of the disturbance cannot be made far enough in advance to permit the system to implement a regulatory response that matches the initial disturbance—i.e., the system is one or more steps behind the environment. Second, the causes of the disturbance cannot be disintangled from the web of interrelations comprising the environment. Both of these problems render anticipatory and reactive forms of regulation ineffective; instead, the system must attempt to simplify the environment. This requires a more active form of control whereby the system acts upon the environment to influence those conditions that are subsequently adapted to. By acting upon the environment, the system attempts to reduce environmental complexity to a more simplified state that can be controlled within the regulatory capacity of the system. Failure to reduce environmental complexity leaves the system with two choices: to leave the environment and move toward more simplified conditions, or to remain and run the risk of being destroyed by uncontrolled forces. Although the choice is never this clear, this discussion underscores the premise that certain environmental factors—rate of change and degree of interconnections—have profound effects on a system's ability to make decisions and to implement effective regulatory responses.

Now that we have examined the necessary conditions for effective regulation and explored some of their implications for system control, we can apply this knowledge to socio-technical systems. The regulation and control of socio-technical systems is similar to the general model of system regulation outlined above. Differences appear, however, when we consider the composition of such systems; for it is here that the independent but correlative nature of the social and technological components is most critical. The system as a whole must not only regulate its overall behavior, but also the performance of each component. Thus, socio-technical regulation occurs at two levels simultaneously, the level of the entire system and the next lower-level of the components.

Returning to the illustration of an open-heart surgical team, the multi-level nature of socio-technical control may be better understood. Each component must maintain its steady state condition if it is to function properly. This requires certain adjustments by members of the social system who are responsible, in this example, for regulating the performance of both social and technical components. At the same time, the steady state for the overall system—in terms of the rate of progress of the operation, for instance—must be maintained. Again, this demands certain control behavior from the team members, although such system-level control could also emanate from an external source, such as the head of the hospital surgical department. This shows the complexity involved in regulating most socio-technical systems. The interactions between the regulatory processes of each component, and between these processes and the regulation of the entire system place heavy demands on the regulatory capacity of such systems.

Like all open systems, socio-technical regulation is concerned with two kinds of steady state variables—those associated with the import and export process and those related to the conversion process. As we have already mentioned, one of the primary functions of the boundary is to regulate the imports and exports required for the performance of the primary task. The boundary performs this function through the use of negative feedback: steady state variables having to do with kinds and rates of import and export, for instance, serve as goals to direct the boundary's regulatory behavior; deviations from these goals inform the boundary of the need for corrective behavior; and the boundary regulator chooses a course of action and implements it to bring the exchange process back to its steady state condition. This regulatory function is frequently performed by members of the social system, but it may also be carried out by a variety of control technologies designed to perform this function—e.g., automatic inspection devices. We must also remember that boundary control operates at both system and component levels, although the composition of the steady state variables is quite different at these two levels. Import and export variables associated with the steady states of the social and technical components are specific to the composition and dynamics of these systems, while exchange variables for the system as a whole are relevant to overall relationships with the environment. Thus, for example, the boundary control process for the social system may involve steady state variables related to informational or perceptual exchanges that have social/psychological meaning for the system members—e.g., "orders to work faster." The same control process for the technological system may concern matter-energy exchanges such as kinds of raw material input. Boundary control for the entire system may involve such variables as rates of output of a finished product.

Regulation of the conversion process of a socio-technical system follows the same general principles as that for the import and export processes.

Here the steady state variables are related to such things as rates and kinds of conversion. The system attempts to reduce deviations from certain ranges of stability required for the effective transformation of imports into exports. In many socio-technical systems, the social component carries out this regulatory function for the technical component. Indeed, much of the social activity which occurs in such systems is aimed at controlling the conversion process of the technical system. Machine adjustments and work flow modifications represent these kinds of behavior.

Since regulatory behavior is directed at reducing deviations from a preferred steady state, all goal-directed behavior in a socio-technical system is a form of regulatory activity. This follows logically from the premise that it is deviations from goals or preferred steady states that direct system behavior and not some blind internal mechanism. Members of many socio-technical systems operate intuitively according to this premise. Forms of behavior that advance the system towards its goals are rewarded, while other kinds of activity—goldbricking, make-work, and other counter-productive behaviors—are punished. When viewed from this perspective, socio-technical systems are goal-directed only to the extent that they are carrying out regulatory behavior toward a preferred steady state. Of course, the issue of which preferred steady state remains a question of perspective. In organization-level systems, the recurrent problem of individual versus organizational goals is only a problem if one set of goals is valued over another. Thus, workers who direct their behavior toward the steady states of their work groups, the Hawthorne workers, for instance—are a problem only to those who prefer different steady states—the Hawthorne managers, in this case.

The choice of a preferred steady state is related to a fundamental problem in the regulation of socio-technical systems: the choice of a basic method of regulation. Emery (1973) discusses this problem in terms of the design of self-regulating systems. He states:

> The choice is between whether a [system] seeks to enhance its chances of survival by strengthening and elaborating special social mechanisms of control or by increasing the adaptiveness of its individual members (p. 71).

Emery's argument is concerned with the redundancy required for self-regulation. A system must have built-in redundancy if it is to have a varied set of responses to deal with a varied set of environmental conditions. This is critical for system regulation as an arithmetical increase in redundancy leads to a logarithmic increase in reliability. A system may achieve redundancy in two ways: redundant parts or redundancy of functions of the individual parts. Each method for attaining redundancy is associated with a different form of system regulation, and each is suited to a particular set of environment conditions.

Socio-technical systems designed with redundant parts employ special mechanisms of control. These control mechanisms—supervisors, for example—determine which parts of the system are active or redundant for a specific regulatory response. Since the control mechanisms must also have redundant parts if they are to be reliable, further mechanisms of control are required and so on. This form of regulation leads to a hierarchy of control mechanisms, each level being controlled from the next higher level. To maintain this form of control, "the system must provide a large number of redundant parts, hence the tendency is toward continual reduction of the functions, and hence cost, of the individual parts" (Emery, 1973, p. 72). Organizations that use a divison of labor work design principle are prime examples of this form of regulation. By dividing work into its simplest units and assigning these to individual workers, the organization attempts to reduce the cost of training and replacing workers. But, since each worker performs a limited function, the organization requires supervisors to control the performance of these functions. Supervisors are able to control only so many workers, hence they perform limited functions which require additional supervision. Eventually, a pyramid of control is constructed where each cog serves a particular purpose. This structure is well-suited to anticipatory and reactive forms of regulation where the system can institute a common set of control programs to cover a large number of redundant parts. These programs inform similar parts, performing common functions, when to respond to a particular disturbance. As we have already discussed, anticipatory and reactive forms of control are effective only when environments are relatively simple—that is, when they have slow rates of change and low degrees of connectedness of parts. Thus, socio-technical systems that employ redundant parts, with resulting pyramids of control, are adaptive to simple environments only.

On the other hand, systems designed with redundancy of functions of parts are suited to more complex environments. Here, the individual parts—workers, for example—possess a range of functions adaptive to a variety of circumstances. Instead of having higher levels of control, the system builds within the individual parts mechanisms for choosing among their various functions a response appropriate for the conditions at hand. In the technological component, this requires multi-purpose technologies with internal mechanisms for switching among different operations as the circumstances demand. More automated forms of technology come close to this requirement, although they ultimately require some degree of social control. For the social system, this form of regulation demands multi-skilled individuals able to coordinate their activities toward a common, preferred steady state or goal. This method of regulation is well-suited to active forms of control where the system members must act upon the environment to reduce its complexity to a more manageable level. Depending upon external conditions, members of the system may deploy themselves in various ways to control a wider variety of forces. Since this can be considered an attempt to match system variety with environmental variety, socio-technical systems with a redundancy of functions of parts are able to cope with complex environments.

In summary, open systems are goal-directed in that their behavior is directed at reducing deviations from a preferred steady state. Such behavior is referred to as regulation and control since the system regulates itself in relation to its steady state. System regulation involves negative feedback, whereby information about the output of the system alters subsequent inputs in the direction of reducing deviations from preferred steady states. Positive feedback, on the other hand, increases deviations from steady states, and although the system requires such feedback for growth, it must somehow control this form of feedback if it is to survive such changes. Effective regulation also demands that the system possess the requisite variety of responses to match the variety of environment disturbances. Known as Ashby's Law of Requisite Variety, this principle states that effective regulation requires both the requisite number and appropriateness of responses for particular disturbances. Even if the system follows this law, it must still be able to choose and to implement a course of action to match environmental conditions. Stable environments—with low rates of change and low degrees of interconnections—allow systems sufficient lag time to implement a response before the initial cause of the disturbance changes. Complex environments—with high rates of change and many interconnections—do not allow systems this advantage; instead, systems must be able to act upon these environments to reduce their complexity to a more manageable level.

Socio-technical systems follow these same regulatory requirements. They differ only by virtue of their independent but correlative social and technical components. Because of these components, such systems must not only regulate their overall behavior, but they must also control the behavior of each of the components as they operate jointly. The interactions among these different regulatory processes place heavy demands on the system's ability to control itself. This is especially salient given the differences between the two components. The social system demands a control process commensurate with its own content and dynamics, while the technical system has its own set of requirements. Steady state variables associated with both import and export processes and with conversion processes display these differences.

The primary issue is to maintain the overall performance of the system while keeping each component in balance. This requires a choice in the design of regulatory mechanisms. Emery has outlined this choice as being between redundancy of parts and redundancy of functions of parts. The former relies on special mechanisms of control and results in a hierarchy of control levels; the latter depends on built-in mechanisms of control and results in a self-regulating system. Each method of control is suited to a different form of environment. A hierarchical control structure—bureaucratic organizations, for instance—matches a simple environment where anticipatory and reactive control mechanisms are effective across a large number of parts. Similarly, a self-regulative control structure matches a complex environment where active control processes are required to deal with a variety of disturbances and to reduce environmental complexity.

Equifinality

Open systems develop towards states of greater complexity and size. In doing this, they display the property of "equifinality" in that they are able to reach a particular steady state from a variety of initial states and in different ways. Because of this property, the steady state of an open system is time independent. The steady state is determined by the system's parameters only—its steady state variables—and not by previous conditions of the system. This contrasts to state determined systems, whose state at any moment in time is determined completely by the previous state of the system. A clock, for example, is state determined, since its state, as measured by time, is determined completely by its previous state.

The property of equifinality is essential for the survival and growth of open systems. Such systems are able to maintain a certain stability and constancy of direction in spite of fluctuations in their environments. They are capable of forming relatively stable, adaptable states because they are not constrained to follow one particular path, but instead can proceed through a series of different steady states, each matched to a particular set of external conditions. This ability to move from one steady state to another by a variety of means allows open systems to continually adapt to environmental changes while keeping their basic integrity or form intact. Indeed, it is this succession of steady states, each standing in a specific relationship to an environment, that we refer to when we speak of the growth or development of a system. A human being, for instance, develops from a fertilized egg into a mature adult by this process. At each stage of development—fetus, infant, child, adolescent, and adult—the organism maintains a particular steady state that is adaptive to the conditions at hand, and depending on these conditions, the organism is able to move to its next stage of development in a multitude of ways. Of course, if the system functions abnormally or if environmental conditions outside the normal adaptive capacity of the system are encountered, the system may cease to develop or perish.

When applied to socio-technical systems, equifinality refers to the ability of such systems to arrive at a preferred steady state from a variety of initial conditions and in different ways. In essence, there is no one best way to design a socio-technical system, but given certain social and technological components, a specific environment, and a preferred steady state to achieve, there is choice in designing the relationship between the components to produce a desired outcome. Thus, the essential question is how to design socio-technical systems so that they are capable of growing toward fully developed states. The answer lies in the dynamics that underlie the developmental processes of the social and technological components.

Technological development involves mechanical construction. Since such systems cannot create themselves, they are designed and implemented by human beings. This process involves mechanical construction, and it is capable of being instituted in one step, from design to structural implementation. Social development, on the other hand, proceeds through a series of growth stages to a fully developed condition. Herbst (1966) summarizes this essential difference:

> Nature does not create in the way factories do. A seed does not contain a complete specification for the organism, and the information given by the genes does not provide for a one step implementation. Yet in spite of the fact that the information given by the gene structure is quite limited, the growth process proceeds with a high degree of reliability and self-maintenance properties at each stage until a viable organism is completed

> which structurally reproduces the original one with a high degree of reliability (p. 14).

Given this fundamental difference in the developmental processes of the social and technological components, it is imperative that we account for both in the design of socio-technical systems. If, like some people advocate, such systems are essentially technological, the problem of system development is rather simple: fully specify the desired state of the system, and design and implement it in one step. For those who use this procedure—division of labor specialists, for instance—problems of one-step implementation frequently arise in respect to the social component. For it is here that growth proceeds in a different manner. "The conditions of both psychological and organizational growth are of the same basic type as biological growth processes as against the mechanical construction type." Further, "a biological organism is not created but it creates itself given an initial structure and a correlated succession of suitable environments which maintain and feed the growth process' (Herbst, 1966, p. 14). Thus, for socio-technical systems containing components that follow fundamentally different developmental processes, it is necessary to account for both forms of growth in the design process.

Again, we are faced with a choice of basic system design. For those systems that employ redundant parts and specialized mechanisms of control, the design process usually follows the mechanical construction type which may be referred to as "complete specification design". The functions of the parts are fully specified in advance, and they are simply joined together like building-blocks to form a completed work structure. The main problem of this approach is that the social component is reduced to the same dimension as the technological component. There is little if any account taken of the growth processes of the social element of the system. A result is that the completed work system frequently fails to operate according to specification, since the two components are out-of-synch, as it were, in regard to their stages of development. In short, the self-maintaining capacity of the system, its tendency toward equifinality, is thwarted by specifying fully the conditions of the system in advance.

The form of system design related to redundancy of functions of parts more fully accounts for differences in growth processes between the two components. In what has come to be referred to as "developmental system design" (Herbst, 1966), the conditions for a fully developed socio-technical system are not specified fully in advance, but rather minimal conditions for effective regulation are built into the system as an initial structure. The system is then allowed to develop itself toward a preferred steady state. These minimal conditions for control correspond to the necessary requirements for effective regulation outlined previously. These conditions enable the system to regulate itself as it develops toward its preferred steady state; they also provide the system with the necessary freedom of response to take full advantage of its property of equifinality.

Again, Herbst (1966) succinctly summarizes our argument:

> If we want, for instance, to implement viable (self-regulating socio-

> technical systems), then the design does not consist of a specification of the final system (although the characteristics which are aimed at [preferred steady states] have to be defined and accepted). What has to be specified and implemented are the conditions for a system of this type to develop; the social system aimed at can rarely be implemented in one step but will need to go through successive stages of growth. The technical design should in this case be such that a workable socio-technical system exists at each stage . . . (p. 14).

SUMMARY

Socio-technical systems theory is based on two fundamental constructs: directive correlation and open system theory. The former provides the basis for conceiving of work systems in which human beings interact with technological components as being composed of two independent but correlative systems—a social system and a technological system—while the latter furnishes a basis for viewing such systems as being interdependent with their environment and hence, open socio-technical systems. As open systems, socio-technical systems display the following properties:

1. An import-conversion-export cycle replenishing the system and permitting it to exist at a high level of complexity.
2. A boundary both separating and relating the system to its environment.
3. A set of steady state variables defining the system and setting parameters for survival and growth.
4. A regulatory process enabling the system to maintain a relatively steady state while engaging with an environment and performing work.
5. An ability to achieve a preferred steady state from a variety of initial states and in different ways.

6

MANAGING SOCIO-TECHNICAL SYSTEMS

A theoretical foundation for understanding socio-technical systems is a necessary prerequisite for managing such systems. The concepts of directive correlation and open system theory direct attention to certain systemic properties of work systems that must be managed effectively if socio-technical systems are to survive and to develop. Presently, there exists no clearly defined calculus for interpreting management practice into these concepts. Rather, the theoretical foundation provides a sound basis for deriving management practice; therefore, management functions congruent with socio-technical theory need to be developed. This should provide a first approximation of what the management of work comprises from this perspective. The chapter is divided into: (1) managing the socio-technical relationship and (2) managing the system and environment relationship.

MANAGING THE SOCIAL AND TECHNICAL RELATIONSHIP

Based on the premise that the performance of a socio-technical system depends upon the relationship between social and technological components, optimal system performance requires that both components be jointly optimized. A jointly optimized work system is one in which the task requirements of the technical system and the biological and social/psychological needs of the social system are both satisfied. This condition must also have economic validity in terms of the financial requirements of the work system or of the larger organizational entity within which it is embedded. Given the requirement of joint optimization, the essential question is how to manage the relationship between the social and technological components to bring about a social and technical match.

The Work Relationship Structure

The answer to this question lies in an understanding of how technical systems are related to social systems. We will use the concept of work relationship structure as a heuristic device to describe the work organization that is brought into existence to relate technology to people. A division of labor work design, for example, is a work organization in which highly prescribed work roles relate the task requirements of mechanized technologies to workers. A work relationship structure specifies the work roles and interrelationships that people must occupy if they are to perform the tasks of the technical component. Since work roles relate individuals to tasks,

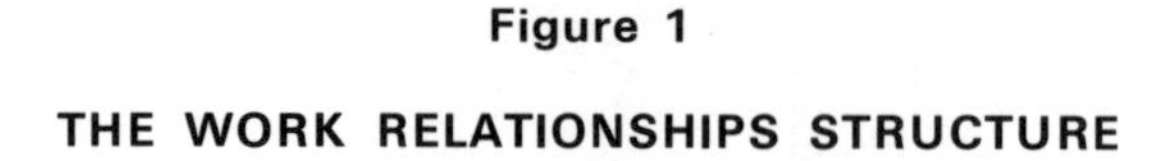

Figure 1

THE WORK RELATIONSHIPS STRUCTURE

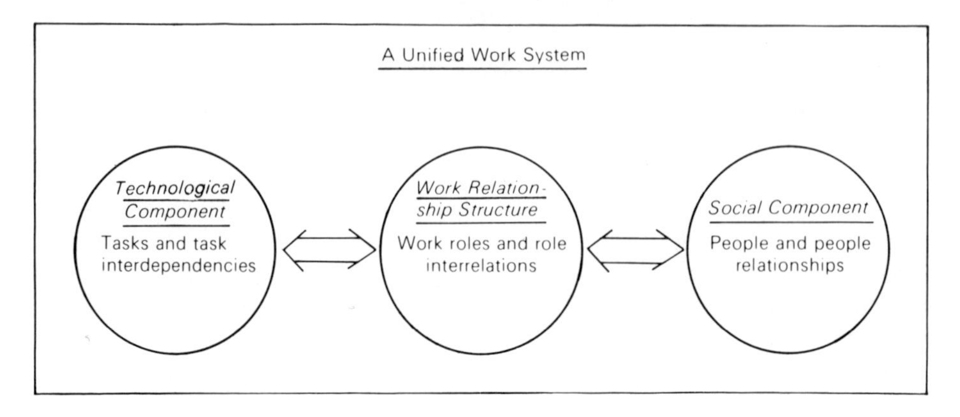

they are the links that tie people to technology: "In one direction they are related to tasks which are related to each other; in the other, to people, who are also related to each other" (Trist and Bamforth, 1951, p. 14). Figure 1 illustrates the way the work relationship structure, by specifying roles and role interrelations, joins the social and technological components into a unified work system.

A work relationship structure provides an organization for relating those who perform the required tasks to each other. Depending upon the specific work roles and their relationships, each work organization constitutes a particular structure for organizing man's relationship to technology. Although the technical component places demands on the kind of work structure possible, the structure itself has social and psychological properties influencing workers. Thus, a jointly optimized work structure must account for both technological and sociological forces. Trist and Bamforth (1951) underscore the importance of both kinds of factors in relation to the long-wall method of coal mining:

> The longwall method will be regarded as a technological system expressive of the prevailing outlook of mass-production engineering and as a social structure consisting of the occupational roles that have been institutionalized in its use. These interactive technological and sociological patterns will be assumed to exist as forces having psychological effects in the life-space of the face-worker, who must either take a role and perform a task in the system they compose or abandon his attempt to work at the coal face (p. 5).

Given the technological and sociological forces existing when individuals occupy work roles and perform tasks, our concern is to design work roles and role interdependencies to fulfill the task requirements of the technical component and the biological and social/psychological needs of workers. Technologically, our focus is on the way tasks are grouped to constitute specific work roles; socially, our interest is on the way work roles present individuals with certain psychological realities that may either thwart or facilitate the fulfillment of their needs. To the extent that work roles are structured so that the technological and sociological forces subsequently emerging result in task performance and human need fulfillment, a jointly optimized work structure is likely to exist.

Since the design and implementation of a work relationship structure is fundamental to our concept of socio-technical system, *a basic function of management is to bring into existence a jointly optimized work relationship structure.* Although the design of work is traditionally ascribed to industrial engineers, managers cannot afford to take the design of work as an initial given from which to start their managerial duties. Managers who relinquish the design of work to others may be so locked into the constraints inherent in a particular structure that they are limited in what they can and cannot do. To overcome such constraints, managers must define their roles to include the design of work. This conceptualization does not reduce the need for the industrial engineering function; indeed, if managers were to define their roles in these terms, they would be better able to engage with industrial engineers in the design of jointly optimized work organizations.

Currently, there exists no laws of joint optimization that would inform managers about how to design work, given a particular technology and a specific group of people. We do know, however, that there is choice in relating technology to people. The problem is to discover the appropriate structure for a specific work setting. This requires detailed knowledge of the technical component and the tasks and task interrelations required for its operation. These task requirements must then be translated into a work organization responsive to both technical demands and workers' needs. The next chapter outlines a strategy for implementing socio-technical systems in organizations. Here we more fully explain the parameters of joint optimization so that a sound basis exists for applying the strategy. We begin with a discussion of the structure or statics of jointly optimized work systems, then proceed to the process or dynamics of joint optimization.

The Structure of Joint Optimization

Structure involves the physical arrangement of a system's components. Our concern is to structure the social and technological components to arrive at a jointly optimized condition. This requires knowledge of tasks and their requirements for performance, and of people and their needs for fulfillment. From this knowledge, it should be possible to hypothesize work structures that are likely to meet both of these requirements.

Technological Considerations

An examination of the technological component and of the tasks and task interdependencies required for its performance is a logical starting point. Our previous discussion of technology revealed that there are a number of technological characteristics appearing to have significant consequences for social systems. These include such factors as the level of mechanization, the physical work setting, and the spatio-temporal layout. The combination of these characteristics for a particular technology places certain task demands on those who operate it. A steel mill under its technical conditions of mechanization, physical setting, and spatio-temporal layout presents workers with certain emergent task requirements that are unique to this kind of technology. Although the various technical characteristics combine to form rather unique work environments, similar technologies share certain features distinguishing them from other technologies. Thus, it is possible to differentiate an assembly-line from a machine shop by virtue of these characteristics.

An initial step in the design of work is to discover how the various technical characteristics combine to form specific task requirements. For a given work system, these technological facts, as it were, may vary in importance among themselves, and a particular characteristic also may vary in saliency over time. A detailed analysis of the technical system should reveal which technical factors or which combination of factors are significant for task determination. Although the next chapter includes a method for analyzing the technical system, one task characteristic appears salient in most work systems. This dimension involves *individual task dependencies*, and we are concerned here with the way such dependencies necessitate certain task groupings and hence the design of work relationship structures.

Emery (1959) states that "the simplest conceptual distinction is between individual tasks that are dependent and those that are independent with respect to their performance" (p. 18). *Independent tasks* do not require cooperation between workers, since they can be completed by one person. Writing a book is an example of this kind of task. Since independent tasks are relatively self-contained, there is considerable freedom in arranging them for task performance. Depending upon their size and complexity, one or more tasks may be assigned to a work role, and a number of these work roles may then be grouped in a variety of ways. One criterion for grouping work roles with independent tasks is to minimize requirements for external control. This arrangement consists of a number of similar work roles assigned to a supervisor. The size of this structure is limited by the number of people one person can manage effectively. Other criteria for grouping independent tasks into coordinative structures are dependencies in the supporting services required to perform such tasks, and dependencies with respect to a common goal. The former dependency requires a work relationship structure that groups work roles according to similar support services such as maintenance and information feedback; the latter demands a structure in which tasks contributing to a common goal are grouped together—e.g., accounting, drafting, and planning tasks are often grouped together because they contribute to a particular sales package. In short, given the freedom to organize independent tasks, our concern is to minimize needs for external control, maximize the delivery of support services, and reduce the distance over which information or materials must travel.

Dependent tasks, on the other hand, require cooperation between workers. This requires task groupings that facilitate the completion of an overall task. Task dependencies may take two forms: "simultaneous interdependence" and "successive interdependence." Emery (1959) states that the former kind of interdependence involves a task that is "too large for an individual to perform in the required time and hence is broken into individual part-tasks" (p. 18). Two individuals operating a double-handled, cross-cut saw represent this kind of task. Successive interdependence involves tasks that must be carried out in sequence Trist et al.'s (1963) description of coal mining in which the performance of one set of tasks precedes the start of another set shows this form of dependence.

Dependent tasks, unlike independent tasks, present certain constraints on the possible work structures that may be designed for their performance. Since the various part-tasks must be coordinated to produce an overall result, the work roles assigned to these tasks must be grouped to maximize coordination around a whole task. A major constraint on coordination is the degree of dependence between part-tasks. This is partly determined by the time lag between the end of one task and the performance of the succeeding task. The longer is this time lag, the greater is the variation in design alternatives. The insertion of buffer inventories between successive tasks is a common method for increasing the time lag between tasks. Another constraint related to the degree of dependence is when more than one work role is assigned to different part-tasks. When sequential tasks are assigned to different roles, problems of overall task coordination may arise if the different tasks are not of equal duration and size. Thus, one worker may perform a task either faster or slower than a related worker, causing discontinuities in the flow of materials or services. The problem of balancing an assembly-line represents this constraint. Given these problems there is still considerable choice in designing work relationship structures to account for task dependencies. The major criterion is to group work roles around a whole task. Then, within this task boundary, problems of time lag and balance delay time may be resolved through the expedient assignment of part-tasks to work roles and the use of different balancing methods, such as buffer inventories, to smooth the flow of materials or services through the successive production stages.

In summary, the technological system presents workers with certain facts or requirements for task performance. The combination of these characteristics constitutes a specific technical milieu for a work system. Attempts to structure work must account for these technical requirements if the primary task is to be carried out effectively. Thus, a first stage in the design of a work relationship structure is to discover how these characteristics combine to form specific tasks. This requires a detailed analysis to find out which factors are significant for task determination in a specific work system.

Although the importance of the various technical characteristics may vary depending upon the work system, there is one factor that is salient in most forms of work. This dimension involves task dependencies, and we are concerned with the way such dependencies necessitate task and work role groupings. Task dependencies may be categorized as either independent or interdependent. The former does not require coordination between workers, while the latter demands coordination between individuals who perform different parts of an overall task. Independent tasks afford considerable freedom in designing work structures; such tasks are usually grouped to minimize external control, to maximize the delivery of support services, and to reduce the distance over which materials and information must travel. Interdependent tasks present certain constraints on work structuring. Such tasks demand structures maximizing coordination in relation to an overall task. Once this is achieved, problems of time lag between part-tasks and balance delay times may be resolved using different balancing methods such as buffer inventories.

Social Considerations

The discussion of the social system in the preceding chapter showed that human beings are social animals who directly and symbolically relate to their environment to fulfill certain needs. This view of man implies that people are motivated to behave when they perceive that such behavior will satisfy their needs. Thus, behavior is directed towards those things that bring satisfaction. Human beings, unlike other animals that do not possess a symbolizing ability, are influenced both by socially and biologically based needs. The latter kind of need, such as food, water, and sex, is endemic to all animals, while the former kind, such as achievement, recognition, and esteem, is unique to social beings who derive such needs from their symbolic relations with one another. Since human behavior is directed toward both kinds of needs, we must account for both in the organization of work. Therefore, understanding how the structure of work affects workers' behavior requires a motivational theory that relates the design of work to the satisfaction of both biological and social/psychological needs.

A current theory of motivation is well-suited for our purpose. Referred to as expectancy theory, this conceptualization explains how the design of work affects workers' behavior or performance (Vroom, 1964; Atkinson, 1964; Porter and Lawler, 1968). For our discussion of motivation, we will use Porter and Lawler's version of expectancy theory. According to their theory, motivation to perform effectively is determinated by two factors: as individual's subjective probability that increased effort towards performance will result in a given reward, and an individual's perceptions that this reward is valued. The first factor, effort-reward probability, consists of two subsidiary concepts: expectancy—a person's belief that effort will result in performance—and instrumentality—an individual's perception that performance will result in reward. The second factor, reward-value, can be explained in terms of an individual's needs; a particular outcome is valued if it satisfies one or more needs. In our case, we are concerned with a person's biological and social/psychological needs in the context of work.

Figure 2 illustrates an expectancy theory of worker motivation. Each of the terms in parenthesis—effort-performance, performance-reward, and value—combine multiplicatively to determine an individual's motivation. Thus, if either is low or nonexistent, then there is no motivation. Consider the case of a worker who values money but feels that increased performance will not result in more money. For him or her, money is not a motivator. Similar arguments can be made for low values on any of the other dimensions that determine motivation. In short, for motivation to exist, a worker must experience high effort-performance and performance-reward probabilities in addition to having a high value for the reward.

Figure 2

AN EXPECTANCY THEORY OF WORKER MOTIVATION

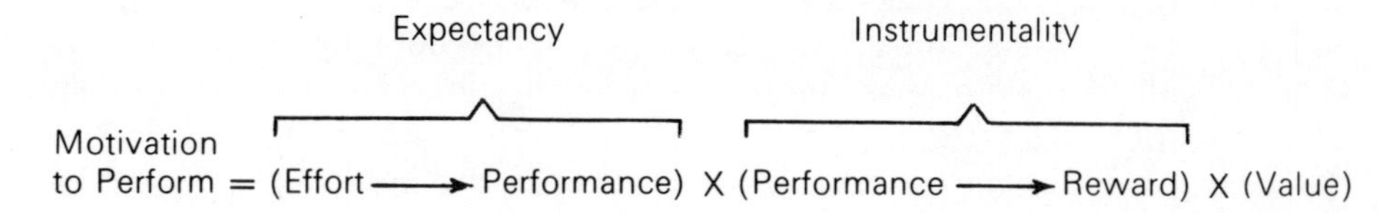

Given expectancy theory as a motivational base, Lawler (1969) explains the relationship between work design and motivation in terms of the performance-reward dimension of this theory. He postulates that work design affects motivation by either increasing or decreasing an individual's perceptions that performance will lead to valued rewards. His argument is based on the difference between extrinsic and intrinsic rewards. Extrinsic rewards are external to the individual, since they are externally mediated by others. Money, for instance, is an extrinsic reward given by others for the performance of tasks. Intrinsic rewards, on the other hand, are internally mediated since the individual rewards himself. Self-esteem and feelings of achievement are examples of this kind of reward. The difference between extrinsic and intrinsic rewards has important implications for work design:

> The fact that (intrinsic) rewards are internally-mediated sets them apart from extrinsic rewards in an important way. It means that the connection between their reception and performance is more direct than is the connection between the reception of externally-mediated rewards and performance. Hence, potentially they can be excellent motivators because higher (performance-reward) probabilities can be established for them than can be established for extrinsic rewards. They also have the advantage that for many people rewards of this nature have a high positive value (pp. 428–429).

Lawler proposes that work design is the critical determinant of whether individuals perceive that good performance leads to intrinsic rewards—this is based on the notion that such rewards derive from the performance of work itself and not from an external source. Thus, it is the specific content or design of work that mediates between performance and feelings of accomplishment and self-esteem. To the extent that work is designed to enhance this relationship, we can expect individuals to be intrinsically motivated. This raises the question: What characteristics of work are more likely to create conditions such that workers expect that good performance will lead to intrinsic rewards?

Lawler identifies three aspects of work conducive to intrinsic motivation. The first aspect concerns feedback about one's performance. Workers must have meaningful feedback to evaluate their performance if they are to feel that their own behavior leads to good performance. This may require a relatively whole task as well as self-defined kinds of feedback. The second characteristic involves the extent to which individuals feel that work challenges their valued abilities. This demands a work structure allowing individuals to employ fully those valued aspects of themselves. The third aspect of intrinsically motivating work entails control. To the extent that workers feel that they have control over setting goals and determining methods for reaching these goals, they are likely to experience that good performance is under their own control and not determined by outside sources. Indeed, each of these work characteristics increases the relationship between a person's feelings of self-worth and good performance by placing the locus of causation for good performance on the individual rather than on the environment. Thus meaningful feedback, methods for testing valued abilities, and work control provide workers with a powerful sense that their own actions cause good performance, and this sense of personal causation is intrinsically rewarding.

The application of expectancy theory to work design raises an important issue about differences in individuals. Since work design is related to intrinsic rewards, we are concerned with the extent to which workers value such rewards. In those instances where workers value intrinsic rewards, work designs providing for meaningful feedback, challenge, and autonomy appear to motivate workers to perform. But what about individuals who do not value intrinsic rewards?

Recent studies on the relationship between workers' attitudes toward their jobs and community characteristics, cultural backgrounds, and personality attributes suggest that some workers are dissatisfied with jobs that are designed for intrinsic motivation (Blood and Hulin, 1967; Wild and Kempner, 1972; Lorsch and Morse, 1974). In other words, jobs that provide for greater responsibility, challenge, and autonomy may have negative effects on workers. These studies explain this negative effect in terms of personality attributes and in reference to the cultural norms of workers.

Lorsch and Morse (1974) present a valuable refinement to Porter and Lawler's (1968) version of expectancy theory. They suggest that a powerful source of motivation is an individual's feeling of competence in mastering his environment, including his job. Furthermore, a person feels competent on the job when it allows him to behave in ways consistent with his personality. Personality, therefore, appears to influence whether the individual expects that performance in the job will provide him with the intrinsic reward of a sense of competence. Specifically, the following personality characteristics appear to determine what form of work is likely to be motivating:

1. An individual's attitude toward authority.
2. An individual's attitude toward being and working alone or in a highly coordinated group.
3. An individual's tolerance for ambiguity.
4. An individual's cognitive complexity.

Lorsch and Morse's (1974) research may be summarized as follows. Workers who prefer dependent authority relations, structured and highly coordinated work patterns, non-ambiguous situations, and non-complex problems feel competent and hence are motivated by work that is more controlling, less autonomous, less ambiguous, and less complex. While workers who prefer the opposite characteristics are motivated by work that is more autonomous, more ambiguous, and more complex. These findings suggest that workers may receive intrinsic rewards—feelings of competence in mastering one's environment—from work designed quite differently. A critical element appears to be personality, and the design of work must be congruent with a worker's personality, at least on the dimensions mentioned above, if it is to provide him with feelings of competence and, hence, motivation to perform.

Blood and Hulin (1967), in reviewing studies of work design, found systematic differences between workers who were satisfied and those who were dissatisfied with jobs structured for intrinsic motivation. They explain these differences in terms of a construct that ranges from identification with middle-class norms to alienation from middle-class norms. Workers who identify with middle-class norms—a belief in the intrinsic value of hard work, a striving for the attainment of responsible positions, and a belief in occupational achievement—are likely to be motivated by jobs that provide challenge, responsibility, and autonomy, while those who are alienated from these norms are not likely to be motivated by such jobs. Blood and Hulin argue that white collar and rural blue collar workers are likely to identify with middle-class norms, while urban blue collar workers are likely to be alienated from such norms. Their reason for using the urban/rural dimension as an index for determining expected alienation of blue collar workers is as follows:

> Blue collar workers living in small towns or rural areas would not be members of a work group large enough to develop and sustain its own work norms and values and would be more likely to be in closer contact with the dominant middle class. On the other hand, blue collar workers living and working in large metropolitan areas would likely be members of a working class population large enough to develop a set of norms particular to that culture. There is no compelling reason to believe that the norms developed by an urban working class subculture would be the same or similar to those of the middle class (p. 52).

Blood and Hulin's (1967) research about the effects of cultural norms on workers' attitudes toward their jobs provides additional insight into the relation between work and motivation. If we can assume that cultural norms affect individual values, then we can understand why identification with or alienation from middle-class norms determines people's responses to work. Identification with middle-class norms such as striving for greater responsibility and achievement is likely to lead individuals to value intrinsic rewards. Thus, we would expect middle-class norms to provide cultural support or sanction for work designed to be intrinsically motivating, i.e., work that provides intrinsic rewards. On the other hand, alienation from middle-class norms would not be expected to provide cultural legitimization for work designed for intrinsic motivation. Of course, we do not mean to imply that workers alienated from middle-class norms do not have their own set of sub-cultural norms. Indeed, such workers are likely to possess norms affecting the kind of work designs valued or legitimized. The problem for those who design work is to discover what particular norms are operating in a specific work situation and to design work congruent with these norms.

Lorsch and Morse's (1974) and Blood and Hulin's (1967) research raises an interesting issue in regard to individual differences: Do the personality attributes and the cultural norms covary? Though there is no data to answer this question directly, the studies seem to support the notion that these variables may indeed relate to one another. Specifically, workers who prefer dependent authority relations, structured and highly coordinated work patterns, non-ambiguous situations, and non-complex problems may also be alienated from middle-class norms. Conversely, those who prefer the opposite conditions may accept such norms. This relationship seems plausible if we consider that personality and culture may be an interactive process resulting in some congruence between cultural norms and values and specific personality attributes (Aronoff, 1967).

Expectancy theory, when combined with the research on individual differences, suggest a contingency approach to work design. A contingency model starts with a determination of what kind of rewards workers value. If individuals value extrinsic rewards, such as money or security, then job design does not appear to be a source of motivation. Rather, motivation is likely to result if extrinsic rewards are contingent upon performance. Thus, the focus should be more on pay and reward systems than on the content of work. On the other hand, if workers value intrinsic rewards, such as feelings of competence in mastering the environment or self-esteem and feelings of achievement, then the design of work appears relevant to motivation. Since individuals' personalities and cultural norms appear to influence the types of work that are intrinsically rewarding, it is necessary to take these individual differences into account in the design of work. For those individuals who prefer authority relations, structured and highly coordinated work patterns, non-ambiguous situations, and non-complex problems, or who are alienated from middle-class norms, work that is less controlling, less autonomous, less ambiguous, and less complex seems to be motivating. Conversely, those who prefer the opposite conditions or who are responsive to middle-class norms seem to be motivated by work that is more autonomous, more ambiguous, and more complex. Returning to Lawler's (1969) properties of intrinsically motivating work—feedback, challenge to valued abilities, and control—these dimensions appear to transcend the individual differences factors. Regardless of personality characteristics or cultural norms, workers must have some degree of feedback, methods for testing valued abilities, and control over work if they are to experience a sense of personal control or mastery over their work. Thus, a concern for these elements must augment the more tailored approaches to work design mentioned above.

In summary, human beings possess both socially and biologically based needs. Since people relate to their environment to fulfill these needs, behavior is directed to those things that satisfy needs. Our concern is how the structure of work affects workers' behavior. This requires a motivational theory that relates the structure of work to the satisfaction of man's needs. Expectancy theory explains how the design of work affects workers' behavior or performance. According to this theory, people are motivated to perform when they perceive that their behavior will result in valued rewards. These rewards may be either extrinsic or intrinsic. The former are externally mediated by others and include such rewards as money and security; the latter are internally mediated by individuals themselves and include such rewards as self-esteem and self-achievement. The structure of work affects motivation by mediating between a person's performance and the attainment of intrinsic rewards—that is, the performance of work itself either thwarts or enhances a person's feelings of self-worth and accomplishment. Those characteristics of work that are likely to increase such feelings include meaningful feedback of performance, methods for challenging valued abilities, and autonomy. These work characteristics provide workers with a powerful sense that their own actions led to good performance, and it is this sense of personal causation that is intrinsically rewarding.

But what about workers who do not value intrinsic rewards? Recent studies suggest that work designed for intrinsic motivation may have negative effects on workers. These studies explain such effects in terms of personality attributes and cultural norms of workers. The personality perspective suggests that workers who prefer dependent authority relations, structured and highly coordinated work patterns, non-ambiguous situations, and non-complex problems are motivated by work that is more controlling, less autonomous, less ambiguous, and less complex. Those workers who prefer the opposite characteristics are motivated by work that is more autonomous, more ambiguous, and more complex. The cultural norms stance suggests that workers who identify with middle-class norms are likely to be motivated by work that is challenging, responsible, and autonomous, while those who are alienated from middle-class norms are not likely to be motivated by such forms of work.

Both perspectives strongly advocate that we must account for workers' personality and cultural norm differences in the design of work. This suggests a contingency approach to the design of work. For those workers who value extrinsic rewards, making such rewards contingent upon performance leads to motivation. On the other hand, the design of work is a powerful source of motivation for those individuals who value intrinsic rewards. Since the types of work that are motivating depend upon individual personalities and cultural norms, the design of work must account for these individual differences.

Structural Propositions for Joint Optimization

Based on the discussion of the structure of joint optimization, we can now present a preliminary set of structural propositions for joint optimization. They represent guidelines for structuring the relationship between the tasks and task interdependencies required for performance and the biological and social/psychological needs of those who perform such tasks. Since these propositions must be linked to the technical and social requirements that exist in an actual work system if they are to be effective, they are meant to serve as guidelines for generating situation-relevant structures for joint optimization.

Emery (1963) has developed a number of propositions for joint optimization at the individual and group levels of work systems. These include:

At the level of the individual:

a. *Optimum variety of tasks within the job.* Too much variety can be inefficient for training and production as well as frustrating for the worker. However, too little can be conducive to boredom or fatigue. The optimum level would be that which allows the operator to take a rest from a high level of attention or effort or a demanding activity while working at another and, conversely, allow him to stretch himself and his capacities after a period of routine activity.
b. *A meaningful pattern of tasks that gives to each job a semblance of a single overall task.* The tasks should be such that although involving different levels of attention, degrees of effort, or kinds of skill, they are interdependent; that is, carrying out one task makes it easier to get on with the next or gives a better end result to the overall task. Given such a pattern, the worker can help to find a method of working suitable to his requirements and can more easily relate his job to that of others.
c. *Optimum length of work cycle.* Too short a cycle means too much finishing and starting; too long a cycle makes it difficult to build up a rhythm of work.
d. *Some scope for setting standards of quantity and quality of production and a suitable feedback of knowledge of results.* Minimum standards generally have to be set by management to determine whether a worker is sufficiently trained, skilled or careful to hold the job. Workers are more likely to accept responsibility for higher standards if they have some freedom in setting them and are more likely to learn from the job if there is feedback. They can neither effectively set standards nor learn if there is not a quick enough feedback of knowledge of results.
e. *The inclusion in the job of some of the auxiliary and preparatory tasks.* The worker cannot and will not accept responsibility for matters outside his control. Insofar as the preceding criteria are met then the inclusion of such "boundary tasks" will extend the scope of the workers' responsibility and make for involvement in the job.
f. *The tasks included in the job should include some degree of care, skill, knowledge or effort that is worthy of respect in the community.*
g. *The job should make some perceivable contribution to the utility of the product for the consumer.*

At group level:

h. *Providing for "interlocking" tasks, job rotation or physical proximity where there is a necessary interdependence of jobs (for technical or psychological reasons).* At a minimum this helps to sustain communication and to create mutual understanding between workers whose tasks are interdependent and thus lessens friction, recriminations and "scape-goating." At best, this procedure will help to create work groups that enforce standards of cooperation and mutual help.

i *Providing for interlocking tasks, job rotation or physical proximity where the individual jobs entail a relatively high degree of stress.* Stress can arise from apparently simple things such as physical activity, concentration, noise or isolation if these persist for long periods. Left to their own devices, people will become habituated but the effects of the stress will tend to be reflected in more mistakes, accidents and the like. Communication with others in a similar plight tends to lessen the strain.

j. *Providing for interlocking tasks, job rotation or physical proximity where the individual jobs do not make an obvious perceivable contribution to the utility of the end product.*

k. *Where a number of jobs are linked together by interlocking tasks or job rotation they should as a group:*
   i. Have some semblance of an overall task which makes a contribution to the utility of the product;
   ii. Have some scope for setting standards and receiving knowledge of results;
   iii. Have some control over the "boundary tasks."

Over extended social and temporal units:

l. *Providing for channels of communication so that the minimum requirements of the workers can be fed into the design of new jobs at an early stage.*

m. *Providing for channels of promotion to foreman rank which are sanctioned by the workers* (pp. 1–2).

Davis (1957b) has also proposed a series of hypotheses for joint optimization. Since many of Davis' hypotheses relate to Emery's (1963) propositions, they are cross-referenced with a letter corresponding to Emery's listing next to appropriate items. Thus, Davis's set of hypotheses both adds to and complements Emery's guides for joint optimization. Davis suggests that joint optimization is more likely to occur when the work content of a job is in the direction of:

1. Putting together tasks that constitute a "meaningful" unit of activity for the worker (b).
2. Increasing the number of tasks in the job to cover a larger part of the process (a).
3. Putting together tasks that are similar in technological content and skill demands (a).
4. Putting together tasks that are sequentially related in the technical process (b).
5. Providing a sequence of tasks or operations [or organization of work] that provides a "meaningful" relationship between jobs (b).
6. Putting together tasks that will include in a job each of the four types of work or activity inherent in productive work, namely: production [processing], auxiliary [supply, tooling], preparatory [set-up], and control [inspection] (e).
7. Putting together tasks that include some final activities in the process or sub-process, or completion activities of a unit, part or product (e).

8. Putting together tasks that permit completion of the part, product or process (b).
9. Arranging the physical facilities and work flow so that the individual can identify the end use made of his work (g).
10. Arranging the physical facilities so that identification and communication with prior and following work stations can take place as a matter of course (h).
11. Putting together tasks and/or geographically locating the work station so as to facilitate involvement of the worker in his immediate work group (h).
12. Putting together tasks so that the job requires increased worker responsibility for rate of output, quality, etc. (d).
13. Putting together tasks so that the job provides the worker with increased control over his job, that is, that enlarges his scope of decision-making, that permits choice of work methods, tools, work rate, etc.
14. Providing or permitting the development of work methods that are "meaningful" to the individual.
15. Putting together tasks that permit the worker to perceive his relationship [the role he plays] to his organizational work unit and to the company as a whole (h).
16. Putting together tasks that permit the worker to perceive the value of his contribution to the organization, to the community, to society (g).
17. Putting together tasks or dividing the product so that the job is perceived as containing some degree of prestige within the organization or community (f).
18. Dividing the product into units which will permit increased identification with the product (g).
19. Putting together tasks that will result in a job requiring maximum utilization of the worker's skills and abilities (f).
20. Dividing the product into the units (parts, components, documents) which are "meaningful" entities to the worker (b).
21. Arranging the physical facilities and designing of communication networks so that feedback on performance quantity and quality and on central information as to production needs takes place automatically and constantly (d).
22. Providing "meaningful" measures of performance to the individual (d).
23. Providing "meaningful" incentives to the individual (pp. 307–308).

Rice (1958 p. 36–39) has also developed propositions for joint optimization. At the individual level, Rice's propositions are similar to Emery's. At the group level, however, Rice adds some important dimensions to Emery's group-level propositions. These include:

1. *If task performance is efficiently organized, then there will be neither fewer, nor more, engaged on a "whole" task than can efficiently perform it.*

If there are too few engaged on "whole" task performance, stress and strain are inevitable, with resulting poor performance and loss of satisfaction. If there are too many, then sooner or later there will arise a need to deny or otherwise to deal with the guilt involved in the mutual investment in inefficiency (p. 36).

2. *A group consisting of the smallest number that can perform a "whole" task and can satisfy the social and psychological needs of its members is, alike from the point of view of task performance and of those performing it, the most satisfactory and efficient group* (p. 36).
3. *The smallest productive and satisfactory group is a pair, and the next most satisfactory a group of from six to twelve, with perhaps the most satisfactory of these a group of eight.*

   With fewer than six members, the importance of individual contributions is so great that the fear of casualties, however caused and however temporary, may be such as to produce disruptive strain. With more than twelve, the complexities of the multiple relationships to be maintained become too great to be carried by every member, and the group tends to split into sub-groups (p. 37).

(It should be noted that Rice's hypothesis for the optimum size of a work group derives from clinical experience where it is assumed that the individual relates to the group through his psychological identification with other members. Later studies have shown primary work groups to be effective with over forty members [Trist and Murray, 1958]. The explanation for this unexpected increase in group size involves the role systems that operate in a work group performing a whole task. In a developed role system, an individual does not have to identify psychologically with other workers to perform a group task. Instead, an individual has to understand only a limited number of related work roles and know the others in terms of their role requirements. Thus, the key determinant in group size is not interaction that results in psychological identification, but interaction that permits role holders to understand and mutually test out each other's roles.)

4. *Group stability is more easily maintained when the range of skills required of group members is such that all members of the group can comprehend all the skills, and without necessarily having, or wanting to have, them, could aspire to their acquisition.*

   The greater the differences in skill between group members (whether the skills are comparable in level or not), the more difficult is communication likely to be between group members. The less likely, therefore, are they to form a cohesive whole (pp. 37–38).
5. *The fewer differences there are in prestige and status within a group, the more likely is the internal structure of a group to be stable and the more likely are its members to accept internal leadership.*

   Inequalities of responsibility and authority in management commands, for example, can lead to considerable difficulties in relationships between colleagues unless the inequalities are accepted and adequately structured into the group organization (p. 38).
6. *The tasks performed by members of a group should be differentiated according to the skills required and the equipment used, and tasks requiring different levels of skill should be hierarchically structured. The task organization should provide opportunities for change in status within the group, and as a corollary, tasks requiring comparable skills should provide some opportunities for interchange of task between members of the group.*

   The greater the value attached to group membership as an end in

> itself, the greater the fear of its loss and, hence, the greater the danger that the group, as a group, will become more important than the task it was created to perform. If this occurs and, as a result, task performance suffers, then the whole group may suffer such loss of satisfaction because of ineffective performance that, even as a group, it may be unable to continue (p. 38).

It is important therefore that:

7. *When members of small work groups become disaffected to the extent that they can no longer fit into their own work group, those disaffected should be able to move to other small work groups engaged on similar tasks* (p. 39).

The Process of Joint Optimization

Process is concerned with change over time. Since both the social and technical components continually undergo change, we must account for the dynamics of these change processes in the management of jointly optimized work structures. Whereas the previous section of this chapter involved the structure or statics of joint optimization, the process or dynamics of joint optimization involves how such structures are modified to adjust to social and technical change. The essential issue is to provide for a method of work design that allows for the continual modification of structure to meet the changing demands of both components. This requires knowledge of social and technological dynamics, and how this knowledge applies to the management of change.

Social Dynamics

Social dynamics involves appreciation. Over time, social systems—people and groups for instance—are able to increase their capacity to experience and to act upon their environment. Since such systems are governed predominately by animate kinds of laws—e.g., biological and psychosocial—this process is organic in that it proceeds by a series of growth stages. Each stage represents a fully functioning system that is able to maintain itself by adjusting to environmental forces. The system, through a continual process of learning and adjustment, is capable of self-generating or elaborating itself to more complex states compatible with a wider range of external conditions. If at each stage the system relates to a suitable environment, it grows and develops; if it does not, it stagnates at a lower level of development or perishes altogether. Thus, social dynamics constitutes a change process whereby the social system appreciates in its capacity for learning and doing.

Technical Dynamics

Technological dynamics, on the other hand, concerns depreciation. Technical systems—machines and production layouts, for example—cannot create themselves; they are designed and implemented by human beings. Since such systems follow inanimate kinds of laws—e.g., physics, electronics, hydraulics—their change process is mechanical in that it proceeds in one stage from design to structural implementation. Once implemented, technical systems are functionally dependent upon people; they do not posses self-generating capacities enabling them to elaborate themselves in relation to external forces. Thus, they cannot, without the help of people, increase their capacities to learn and to act. Given these characteristics, technical systems depreciate with age. The seeds of depreciation are sown by the very social systems that create and use technologies. As people appreciate in their capacities for learning and doing, they outgrow existing technologies. Since technology cannot change itself to match these new appreciations, it is doomed to obsolescence. In this sense, obsolescence is not built into technical systems, but it derives from the people who outgrow them. Of course, new technologies are invented to replace existing ones, but from the perspective of a specific technological system, change involves an aging process whereby technology depreciates in the hands of those who use it.

Differences Between Social and Technical Dynamics

Differences between social and technical dynamics suggest two factors relevant for bringing into existence work relationship structures accounting for both kinds of change: the rate of change of the components and knowledge of fully developed work conditions.

Generally speaking, the rate of social change is slower than the rate of technological change. This derives from differences between organic and mechanical forms of growth. Organic systems proceed in growth stages toward fully developed conditions. For people, this involves learning new ways to think and to behave at each stage of development. Since this process requires a person to reorient himself perceptually to new conditions, it may be thought of as a continual series of learning-unlearning-learning events. Given the complexity and evolving nature of social change processes, they are usually slower than technical change. The faster pace of technical change derives from its mechanical nature—that is, mechanical systems are of the construction type in which it is possible to proceed in one step, from design to structural implementation. Thus, a machine or production process can proceed from design to a working stage in a rather short period of time, especially in comparison with social change.

Since we are concerned with structuring the relationship between social and technological components, we must account for differential rates of change in designing work structures. Failure to realize that social change is slower than technical change frequently leads to a work structure that is geared to the faster-paced, technical component. A common fallacy is to assume that the work system is basically technological and then to design and implement a work structure in a one-step, mechanical manner. This assumption often results in a work structure that is not synchronized with regard to the rates of change of the components. Thus, the social system, in developing slower than the technical system, is not able to perform effectively. Such problems are often blamed on workers' resistances to change rather than on a realistic appraisal of the time required for workers to develop effective work roles and role relationships to perform the tasks of the technical system. Similar problems may occur when technological changes are introduced into existing work structures. To the extent that such changes are not compatible with the existing work structure, workers must reformulate their work roles to meet new task demands. The amount of time required is usually longer than anticipated. Given the slower rate of social change, we must take this as a limiting factor in the design and implementation of work relationship structures. Thus, the time taken to develop a technological system must also include the time taken to develop a requisite social system to operate it.

In managing work systems, we frequently make assumptions about how the system should operate in a fully developed condition. These assumptions, whether they be valid or false, influence how we design and manage such systems. Technologically, our knowledge of work system behavior is relatively clear. Since the laws that govern the dynamics of technical systems are relatively known, we are able to predict with a high degree of accuracy how technologies should function. Thus, we usually have a set of structural specifications theoretically stating what the technical system can do in a developed condition. For social systems, however, our knowledge of dynamics is not so clear. At best, we have an adequate understanding of the biological and physical laws that govern a person's perceptual and motor behaviors. This allows us to determine optimal work rates and task allocations for relatively simple behaviors, but not for more complex ones. When we are concerned about social systems in terms such as creativity, complex problem-solving, motivation, and social interaction, we are relatively unaware of the optimal conditions for social development. Since our knowledge of these forms of behavior is not well formulated, we often opt for simplistic work structures accounting for only well-known aspects of social behavior and hope that the other parts will somehow sort themselves out. These work structures are usually completely specified in terms of technological and social "knowns," thus placing a constraint on other dynamic factors that may or may not emerge under such conditions. This results in work structures that are geared to the dynamics of technical development to the exclusion of the full range of social dynamics. Since the work system as a whole must account for both kinds of dynamics if it is to function effectively, we are often left with a suboptimal design.

Given differences in the rate of change and our knowledge of fully developed conditions for the social and technological components, management must provide for a work relationship structure allowing the components to develop jointly in terms of their respective dynamics. Since the components must interact to produce a specified result, such structures must account for interaction effects such that changes in one component do not adversely affect changes in the other component. Current methods of organizing this relationship—division of labor principles, for instance—frequently employ mechanical construction kinds of designs. These may be referred to as "complete specification designs" in that work roles and their interrelations are fully specified in advance and the resulting work relationship structure is then implemented in one step. A completely specified work structure does not allow the social system to develop beyond the constraints imposed by the initial design. Thus, social development is compromised at too low a level. Similarly, a work relationship structure that is implemented in one step does not provide the social system with the necessary time to grow toward a fully developed condition. Because of the problems inherent in complete specification designs, we must look elsewhere for a method of work organization congruent with the dynamics of both the social and technical components.

Developmental System Design

The discussion of equifinality in the previous chapter suggests a method of work design that accounts for the dynamics of joint optimization. Referred to as "developmental system design" (Herbst, 1966), this method was introduced for the design of self-maintaining, socio-technical systems. It was argued that the implementation of a jointly optimized work structure does not require a complete specification of the final structure in advance; rather, we must specify the minimal conditions for such structures to develop themselves. Thus, given initial conditions for self-regulation, socio-technical systems are able to elaborate themselves structurally, through a series of growth stages, to more fully developed conditions.

The self-generating capacity of socio-technical systems derives from their property of equifinality—that is, they are able to reach a specified end state from a variety of initial conditions and in different ways. Based on this property, developmental system design suggests that there is no one-best-way to structure socio-technical systems. But, given a specified task, a particular combination of people and technology, and a set of environmental conditions, there are a variety of ways to organize work to produce similar results. Since both social and technical components are continually undergoing change, this method of work design involves a series of work structures, each geared to the particular dynamics of the system. The essential task is to provide an initial self-regulative structure, and then to leave the remaining conditions free to vary in accordance with the dynamics of both components. These free conditions—work role relationships, for instance—provide the system with the necessary flexibility to develop work structures compatible with the emerging dynamics of the system, thus taking full advantage of the system's capacity for equifinality.

Developmental system design requires managers to implement an initial self-regulative work structure, and then to provide for a succession of suitable environments that permit this structure to develop according to its own dynamics. Since we discuss this latter requirement in the next section, we will now outline the necessary conditions for an initial, self-regulative work organization. Herbst (1966) refers to such work organizations as "autonomous work groups," and he identifies the following conditions as minimal requirements for their implementation:

1. The unit should have a clearly definable and easily measurable outcome state which will generally be in the form of quantity and quality of a product, and also an easily measurable set of relevant import states. This provides the necessary information to management for performance evaluation of the system and the necessary information for internal process maintenance and adjustment;

2. The unit should contain all the functions required for process control, maintenance and adjustment;
3. A single social unit is responsible for the total production unit. It must contain all the required technical skills and be capable itself of self-maintenance and adjustment; and
4. Given that the functional elements of the production process are interdependent with respect to the achievement of the outcome state, the social organization should be such that individual members (in the case of a work group) do not establish primary commitment to any part function, that is, do not lay claim to or force others to accept ownership or preferential access to any task or equipment but are jointly committed to optimizing the functioning of the unit with respect to the outcome state as a primary focal goal (pp. 10–11).

The conditions for self-maintaining work systems can be satisfied by many of the structural propositions for joint optimization presented in the previous section of this chapter. Implementing these propositions at either the individual or group levels of work should provide for meaningful feedback, requisite task variety, whole task responsibility, and primary goal commitment. Since the structural propositions are concerned with the static arrangement of a system's components, they do not account for the dynamic processes by which such structures are changed or modified over time. Thus, the propositions provide the structural prerequisites for adaptive work systems, but they do not specify how such adaptation is to occur.

To account for structural change, developmental system design should be considered a tuning process whereby the self-regulative work structure is continuously tuned or adjusted to keep the system operating effectively as the components undergo change; for example, workers may continually adjust the technology or their own work role relationships to maintain a jointly optimized condition. This tuning process will be effective to the extent that workers perceive it to be both necessary for task accomplishment and a legitimate part of their work roles. When workers perceive that both assumptions are valid, they are likely to adjust their work system to changing conditions if the structural prerequisites for self-regulation (outlined above) are present. This should allow managers to free themselves from many of the problems of internal control and focus their efforts toward the more relevant task of relating the work system to its environment. In this respect, developmental system design is a necessary step to managing open, socio-technical systems.

In summary, the management of a jointly optimized work system concerns both social and technological dynamics. The essential problem is to account for the dynamics of both components as they relate to task performance. Socially, this involves an appreciative process in which the social system appreciates in its capacity for learning and doing. Technically, this entails a depreciative process in which the technical component depreciates in the hands of those who use it. Given these differences, the social component changes at a slower rate than the technical component; knowledge of a fully developed condition is also weighted in favor of the faster paced technological system. A jointly optimized work structure must allow for these differences so that the dynamics of one component do not interfere with the dynamics of the other component in the performance of the primary task. Traditional methods of complete specification design do not fulfill this requirement. They tend to constrain the full range of social development and do not provide the social system with the necessary time to grow to a fully developed condition. A method of work system design that accounts for the dynamics of both components is needed. Developmental system design provides for the dynamics of a jointly optimized work structure. By specifying the minimal conditions for a self-maintaining work system, this method of design provides managers with a powerful strategy for engaging workers in the process of jointly optimizing their work structures. An initial self-regulating structure provides workers with the structural prerequisites to continually tune or adjust both social and technical components for task accomplishment. To the extent that workers perceive this tuning process as both necessary for task performance and a legitimate part of their work roles, managers are freed from matters of internal control and are able to attend to the more urgent task of relating the work system to its environment.

Summary

A basic premise of socio-technical theory is that work system performance depends upon the relationship between both social and technical components of work. Specifically, optimal system performance requires that both components be jointly optimized such that the task requirements of the technical system and the biological and social/psychological needs of the social system are jointly satisfied. This premise, when interpreted into management practice, implies that *a basic function of management is to bring into existence a jointly optimized work organization.* This requires the design and implementation of a work relationship structure that ties people and their relationships to tasks and their interdependencies through the specification of work roles and work role relationships. Currently, there exists no laws of joint optimization that would inform managers how to design such structures, given a particular technology and a specific group of people. The next chapter, however, outlines a strategy for implementing socio-technical systems in organizations. Before managers can apply this strategy, they must understand the parameters of joint optimization if they are to effectively implement jointly optimized work structures. This requires knowledge of both the structure and process of joint optimization.

The structure of joint optimization involves the physical arrangement of a system's components. Our concern is to structure the social and technical components to arrive at a jointly optimized condition. This requires knowledge of both tasks and their requirements for performance and of people and their needs for fulfillment. Information about task derives from a detailed analysis of the technological system to discover how various technical characteristics —e.g., the level of mechanization, and the physical work setting—combine to determine emergent task requirements. Although the next chapter presents a method for analyzing the technical system, the task characteristic concerned with dependency of tasks is salient in most work systems. Thus, we must account for different forms of task dependencies in the structure of work. Knowledge about workers and their needs for fulfillment also requires a detailed analysis of the social system to discover what needs are salient in a particular work situation. An understanding of how the structure of work affects workers' behavior requires a motivational theory that relates the design of work to the satisfaction of both biological and social/psychological needs. Expectancy theory explains this relationship. Briefly stated, individuals are motivated to perform effectively when they perceive that their behavior will result in valued rewards. Work design affects motivation by mediating between a person's behavior and internally mediated or intrinsic rewards. Work that is designed for meaningful feedback, challenge, and autonomy is likely to be intrinsically motivating, since it provides individuals with intrinsic rewards through the performance of work itself. Such work designs will not motivate all people, however. We must also account for individual differences in personalities and cultural norms in the structure of work. This requires a contingency approach to work design. Based on the discussion of the structure of joint optimization, it is possible to derive a number of structural propositions for joint optimization at both the individual and group levels of work. These propositions may be used as guides for generating situation-relevant work structures in organizations.

The process of joint optimization involves change in both the social and technological components. Our concern is to account for the dynamics of both components as they relate to task performance. Social change involves an appreciative process, while technical change entails a depreciative process. The former process is slower and less understood than the latter. A jointly optimized work structure must account for these differences so that dynamic interactions between the components do not interfere with the performance of the primary task. This requires a method of work design that provides for the dynamics of both components. Developmental system design fulfills this purpose by specifying the minimal conditions for a self-regulative work structure and leaving the remaining conditions free, as it were, to vary in accordance with the work system's dynamics. This method of work design, when combined with the structural propositions for joint optimization, permits workers to tune or adjust continually both social and technical components for task accomplishment; this frees managers from the task of internal system control and allows them to focus on the more important matter of relating the work system to its environment.

MANAGING THE SYSTEM AND ENVIRONMENT RELATIONSHIP

Socio-technical systems, like all open systems, must interact with their environment if they are to survive. This system and environment relationship may be considered an exchange process; the system exchanges various forms of energy and information with its environment in order to gain resources to survive and grow. To the extent that the system obtains needed resources and furnishes the environment with valued products or services, the system may be said to match its environment. This system and environment match is a necessary condition for effective system performance. Therefore, *a second function of management is to manage the system and environment relationship to assure a compatible match.*

An understanding of this function lies in the knowledge of open systems already presented. Open system properties explain the apparent paradox of how such systems maintain themselves in relatively independent states while interacting with their environment. If this knowledge is to be of use in managing system and environment exchanges, it must be translated into management practice. Since there are no clearly defined principles for open systems management, we are again faced with the task of deriving management practice from the theoretical foundation for understanding socio-technical systems.

Given the premise that socio-technical systems must relate to their environment if they are to survive, a primary function of management is to assure that this relationship is carried out effectively. This management function involves two related tasks. The first task concerns the work system's boundary. Since work systems relate to their environment through a boundary that both differentiates the system and mediates environmental exchanges, it is necessary to enact a boundary such that the system's primary task is separated or protected from external disruptions while the system regulates its exchanges with the environment. We refer to this process as "boundary management" and include it as an essential task of management. The second task involves the management of forces external to the system. Since work systems must relate to a succession of suitable environments if they are to survive and develop, it is necessary to plan for a desired environment and to initiate steps to bring it about. We refer to this second process as "open system planning." When both tasks of management are performed effectively, socio-technical systems are able to perform their primary tasks while managing their relationships with a succession of favorable environments.

Boundary Management

The boundary of a socio-technical system represents both a discontinuity of the system's primary task and an interpolation of a region of control. The discontinuity of the primary task differentiates or protects the internal functioning of the work system from external disruptions, while the region of control regulates exchanges with the environment. Both functions of the boundary must be managed effectively if the work system is to perform its primary task and relate to its environment. Therefore, we shall discuss boundary management in terms of protecting the primary task and regulating environmental exchanges.

Protecting the Primary Task

Socio-technical systems at all levels—e.g., the individual, the group, the organization—have, at any given time, one task which is primary. The primary task is that conversion process which the work system must perform if it is to survive. To perform the primary task effectively, the system must have a certain degree of certainty or control over those variables that affect task accomplishment. One way to achieve this certainty is to seal off or protect the primary task from external forces that may disrupt its performance. Protecting the primary task from environmental influences provides the work system with a boundary of rationality within which the task can be performed with relative certainty. Given this requirement for technological rationality, a major function of boundary management is to ensure that the primary task is protected from external influences.

A first step in protecting the primary task from environmental disruptions is to enact a boundary such that the primary task is clearly differentiated from other activities outside the work system. Unless the primary task is differentiated, there is no way in which activities carried out within the system can be insulated from activities outside. Miller (1959) has suggested three criteria for differentiating primary tasks: technology, territory, or time. Each criterion, or some combination of them, may be used to bound a work system. Technology implies a discontinuity in a conversion process such that similar conversions are grouped into a single work system. For instance, a series of machining processes may be combined to form a machining department. Territory involves a differentiation in the spatial arrangement of task components; for example, the men and machines that share a common physical location may be grouped into a work system. Time refers to a discontinuity in the time span of task performance. Workers who perform tasks on the first shift, for instance, may be bounded into a temporally defined work group. Each of these dimensions provides a useful criterion for differentiating the primary task from its environment. Since the technological criterion relates directly to the primary task definition, it appears to be an initial consideration for task differentiation. The other criteria—territory and time—may be used to bound further the primary task. Thus, similar technological processes may be combined initially into a single work system; then, this work system may be differentiated further into groupings of similar territory or time.

Once the primary task has been differentiated from the environment, other methods for protecting it from external influences may be instituted. Thompson (1967) has identified our strategies for providing such protection: buffering, leveling, anticipating, and rationing. Although he presents these strategies for organization-level work systems, they appear to be relevant for socio-technical systems at all levels.

1. Buffering involves the placement of input and output components at the boundary of the primary task. These components are meant to provide the conversation process with both a continuous supply of inputs and a stable market for outputs. Input buffers—stockpiles of raw materials, for instance—absorb environmental fluctuations before they enter the work system. These irregularities are stored in the input component and then inserted steadily into the system. Thus, the primary task is protected from unnecessary disruptions. Output buffers—finished product inventories, for example—reduce environmental fluctuations originating in the work system's market. By storing the work system's outputs and distributing them in accordance with environmental demands, buffer outputs permit the primary task to operate at a constant rate, free from changing market conditions.

Although buffering is an effective means of protecting the primary task from environmental fluctuations, it is costly to maintain input and output inventories adequate to meet task requirements. Therefore, work systems employing this strategy must compromise between the high costs of buffering and the amount of certainty required for task performance.

2. Leveling is a strategy that seeks to reduce fluctuations in the environment by smoothing out input and output exchanges. In contrast to buffering which absorbs environmental irregularities, leveling attempts to reduce such changes. Leveling input transactions is illustrated by the maintenance of alternative sources of raw materials to ensure a steady supply. An example of output leveling is the inducements, such as discounts, that are offered during slack sales periods. Regardless of the method used to level environmental exchanges, complete smoothing of inputs and outputs is difficult because the environment is, by definition, not fully under the control of the work system.

3. The strategy of anticipating is concerned with protecting the primary task by anticipating and adapting to environmental conditions. This strategy is employed in those circumstances where external forces cannot be adequately buffered or leveled. The objective of anticipating is to forecast environmental fluctuations and to treat this information as a constraint to which the primary task must adapt. Thus, work systems may learn that certain environmental changes are patterned—e.g., seasonal buying habits—and they can use this data to plan for varying rates of task performance. Examples of this strategy are the seasonal hiring practices of the Postal Service and the National Park Service. Anticipating is an effective method for protecting the primary task when the environment is relatively predictable. Simple environments with slow rates of change and low degrees of interconnections are compatible with this strategy, since they permit work systems to anticipate and adapt to their changes. Complex environments, however, are more dynamic and hence, unpredictable. Therefore, anticipating and adapting to complex environments is a difficult if not impossible task.

4. The final strategy, rationing, involves the allocation of primary task performance to selected exchanges. Rationing is usually employed when the other strategies do not protect the primary task. This results when the primary task is not able to transform all available inputs or when the demand for the system's products or services exceeds task performance. A typical illustration of this strategy is the differential allocation of medical services during a community disaster. Since work systems must resort to rationing when their primary tasks are unable to perform at a rate acceptable to the situation, the application of this strategy implies that either the task is not being performed effectively or the task is too limited for maximal exchanges with the environment. The former implication means that rationing is a defensive strategy for protecting an ineffective task, while the latter signifies that rationing is a stop-gap solution for a task with limited capacity. In either instance, rationing should be considered an emergency solution, applied either under unusual environmental circumstances or until the primary task is modified for increased performance.

Thompson's (1967) four strategies for protecting the primary task from environmental changes appear to form a continuum of effectiveness. Buffering is most effective, since, in absorbing fluctuations, the work system can exert considerable control over inputs and outputs. Leveling is next in effectiveness, since the work system has only moderate control over influencing environmental irregularities. Anticipating is less effective because in reacting to an anticipated future, the system has little control over what actually transpires. Finally, rationing is least effective, since it implies an imbalance between task performance and environmental conditions. If this assumption about the differential effects of Thompson's strategies is valid, boundary managers should seek to apply the various strategies in order of effectiveness. To maximally protect the primary task, managers should first buffer it with input and output components. Those fluctuations not effectively controlled by this method should then be subject to a leveling strategy and so on. Obviously, rationing is used only for short-term emergency situations. When Thompson's strategies are applied judiciously, we can expect the primary task to be sufficiently protected from external influences.

In summary, a major function of boundary management is to protect the work system's primary task from environmental disruptions that may affect its performance. This protection provides the work system with the certainty or control necessary for task performance. A first step in protecting the primary task is to enact a boundary such that the primary task is clearly differentiated from other activities outside the work system. Miller's (1959) dimensions of technology, territory, and time are criteria for enacting this boundary. Once the primary task is differentiated, other methods for protecting the primary task may be instituted. Thompson's (1967) strategies of buffering, leveling, anticipating, and rationing are effective methods for protecting the primary task from environmental fluctuations.

Regulating Environmental Exchanges

Protecting the primary task from environmental influences provides the work system with a certain freedom or independence from its environment. This protection, however, is a necessary but not sufficient condition for task effectiveness. In addition to being relatively independent, sociotechnical systems must also exchange forms of energy and information with their environment if they are to survive and to develop. To carry out such exchanges, work systems must regulate their imports and exports such that required materials and information are allowed to enter and leave the system while unnecessary ones are not. Since this regulatory activity occurs at the boundary of a work system, a second function of boundary management is to ensure that environmental relationships are controlled.

An initial task of providing for effective exchanges is to implement boundary control components or activities at those locations where the work system is differentiated from its environment. If environmental relationships are controlled at points of system discontinuity, few problems of boundary confusion are likely to occur. For instance, the placement of an inspection function—a boundary control activity—at the boundary between the work system and its environment is most effective when the boundary clearly differentiates the system from its environment. Since we have already discussed the need for boundaries to protect the primary task from external disruptions, it is apparent that boundaries that protect the work system are also likely to provide points of differentiation required for the control of environmental exchanges. Thus, boundary protection is a prerequisite for boundary regulation; the former provides the differentiation required for the latter.

The placement of boundary control components at both the import and export sides of the work system provides a region of control between the primary task and its environment. This region of control operates in two directions: inward to the internal activities of the system and outward to the environment. Given these two directions, the region of control—in the form of a boundary control component—carries outs its own import-conversion-export process. Using the example of an inspection unit of a manufacturing organization, the import activities are the collection of information about the state of incoming or outgoing materials; conversion activities are the comparison of this information against a standard of acceptability, and export activities are the decisions to allow materials to enter or leave the system. Since these regulatory activities are similar to those carried out by all open systems, we can expect boundary control to require conditions for success similar to those outlined for open systems in the previous chapter. Therefore, in addition to placing boundary control units at points of system discontinuity, managers must also ensure that these components have the following characteristics for successful regulation:

1. A set of steady state variables and their ranges of stability serving as standards for proper system functioning. These variables—import and export variables in this instance—are the regulator's objectives, and they serve as standards of acceptability for judging if certain materials or information may enter or leave the work system. Examples of these variables include: rates, kinds, and costs of imports and exports.

2. A method for obtaining information as to the actual state of these steady state variables for comparisons against the standards of performance. Boundary control units must be able to test the condition or state of imports and exports to ascertain if they meet standards of acceptability. Such tests require observational techniques that yield information about imports and exports in terms of standards of acceptability. Illustrations are automatic sensing devices, accounting procedures, and simple mechanical measuring tools such as templates.

3. A repertoire of behaviors that may be employed to correct for any deviations from the standards found. Boundary regulators must have a variety of corrective behaviors from which to choose an appropriate response, when imports and exports do not meet standards of acceptability. These responses include such corrective strategies as rejecting imports or exports, altering them to meet standards, or varying their rate of flow.

4. A method of decision-making and acting that enables the system to choose and enact a corrective course of action before the cause of the

disturbance changes. This requirement implies that boundary control components must not only be able to choose and to implement a corrective response, but also if this response is to be effective, it must be implemented before the disturbance for which it was intended changes. Thus, for instance, an inspection unit that decides to reject certain raw materials must carry out this corrective response before they enter the system.

The above requirements for effective regulation provide boundary control units with the capabilities necessary to perform their control function. Their application varies depending upon the specific imports and exports of the work system as well as the particular standards of acceptability employed for boundary regulation. When these requirements are implemented in boundary control components located at points of system differentiation, work systems are able to regulate their exchange with their environment.

Summary

The boundary of a work system represents both a discontinuity of the primary task and an interpolation of a region of control. To effectively manage this boundary, the primary task must be protected from external disruptions while the system regulates its exchanges with the environment. Protecting the primary task from environmental influences provides the certainty required for task performance. To ensure such protection, the primary task must be differentiated from its environment. Miller's (1959) criteria of technology, territory, and time may be used to enact this differentiation. Once the primary task is differentiated, other strategies for protecting it may be employed. Thompson's (1967) notions of buffering, leveling, anticipating, and rationing are four such strategies. In contrast to boundary protection, regulating environmental exchanges concerns the control of imports and exports required for system survival and development. Effective boundary regulation requires that boundary control components be placed at points of system discontinuity. Thus, boundary protection provides the differentiation required for boundary regulation. Since boundary control is similar to open system regulation, boundary control units must have the same characteristics for successful regulation as those for open systems.

Open-Systems Planning

Socio-technical systems must relate to a succession of suitable environments if they are to survive and to develop. This requirement implies that work systems must not only be able to exchange with their environment, but the environment itself must also be favorable for system performance. Whereas boundary management provides the system with the capabilities necessary to relate to its environment, it does not ensure that the environment is suitable for the system. This latter problem suggests that an additional requirement for managing system and environment relationships is to influence the environment in directions that are appropriate for the work system. We refer to this management function as "open-systems planning." This concept represents a method whereby work systems can affect environmental realities in directions favorable for system survival and development

(Clark and Krone, 1972). Since this approach is aimed at the environment as an independent entity apart from the work system, it directs attention outward to those forces that affect system performance.

Open-systems planning is based on assumptions about how individuals conceptualize and act upon their environment and about the condition of the environment itself. We discuss both assumptions before applying this method to the management of system and environment relationships.

Assumptions of Open-Systems Planning

Open-systems planning derives from a concern for how individuals perceive and act upon their environment. Using Vicker's (1965) concept of "appreciative system," Clark and Krone (1972) describe this process as an act of appreciation whereby the individual makes certain judgments about external realities and then comes to value these judgments in certain ways. These appreciations, made up of reality and value judgments, provide the individual with a perceptual basis for planning and implementing specific courses of behavior. For example, an individual may appreciate his environment as being composed of predominately middle-class whites which he values highly; this appreciation may guide the individual's behavior in certain directions which may take the form of racism. When our attention is directed to social systems comprised of more than one individual, acts of appreciation involve communication between people—that is, to function as a social system, individuals must communicate with one another if they are to arrive at a shared appreciation of their environment. Shared appreciations are necessary if group members are to generate a sufficiently common view of reality to formulate and to implement a plan to achieve a shared objective.

When applied to work systems, shared appreciations provide group members with a common view of their environment. This view distinguishes certain aspects of the environment as important for task performance and other aspects as relatively unimportant. Based on these appreciations, work systems plan certain courses of action in respect to the environment, and then implement these activities to accomplish their goals. To the extent that members can effectively communicate their individual appreciations to one another, they should be able to arrive at a sufficient consensus of the situation to allow them to plan and to execute a strategy for relating to their environment. This strategy will be effective only if the shared appreciations account for those environmental realities affecting task performance. Thus, we are concerned with generating shared appreciations that result in appropriate courses of action.

This latter concern for matching environment realities leads to a second assumption of open-systems planning: that the state or condition of the environment has a significant effect on whether or not appreciations will be appropriate to the environment. The discussion about the state of the environment—simple or complex—in the previous chapter provides a basis for understanding this assumption. Simple environments are relatively stable; they permit system members to anticipate and to react to external forces with a high probability of success. Complex environments are continually changing in unpredictable ways; they render reactive modes of action obsolete to changing conditions.

These differences, between simple and complex kinds of environments, suggest that appreciations must also differ on similar dimensions if they are to match both types of environments. Thus, individual or group appreciations that are relatively stable and that account for limited parts of the environment are suited to simple environments. These appreciations allow system members to plan and to initiate courses of action that are likely to be effective over long time periods; such plans are also likely to encounter few unanticipated disruptions, since the environment is not likely to change unexpectedly. Similarly, individual or group appreciations that are relatively dynamic and that account for wider aspects of the environment are appropriate for complex environments. The ability to develop new appreciations based on a variety of external data enables system members to continually modify or change their plans depending upon emergent environmental conditions. Such plans are likely to be short-term and adaptive to new circumstances as new appreciations inform individuals about changes in the environment.

When taken together, both assumptions—about how individuals appreciate and act upon their environment and about the state of the environment—allow us to draw some important inferences about how work systems can increase their capacity to relate to their environment. The first inference concerns the need for effective communication among system members if a sufficiently agreed upon view of the environment is to emerge. As work systems encounter more complex environments, the requirement for effective communication increases rapidly. Members must exchange varied kinds of information in a short period of time if they are to develop new appreciations rapidly enough to match changing conditions. This requires open and direct forms of communication among individuals. System members must be able to perceive and value one another's appreciations and feelings; they must account for individual differences and effectively resolve discrepancies into a sufficiently common view of reality to permit action toward a shared perception of the environment. Thus, well-developed communication skills are a necessary requirement for work systems faced with complex environments.

A second inference involves the amount and variety of information that must be gathered to formulate plans for action. As work systems experience more complex kinds of environments, the data required for planning expands quickly. Since the interconnections among parts of the environment are great, individuals must gather information from varied sources if they are to understand the complex dynamics operating. This requires knowledge of the primary and secondary effects of anticipated courses of action. For instance, detergent manufacturers must not only know that their products will get clothes "whiter than white," but also the possibility that secondary effects, such as water pollution, might lead to negative customer reactions. To gather such data, work systems must probe a variety of environmental domains—e.g., political, economic, social and physical spheres of the environment; they must then organize this data to make valid inferences upon which to base their plans and future actions. The greater the complexity of the environment, the greater is the need for work system members to have well developed communication skills and increased amounts and variety of external information if they are to relate to such environments.

Based on the above assumptions about open-systems planning, Clark and Krone (1972) have developed a strategy that enables work system members systematically to learn about their environment and to make plans to influence the environment in desired directions. Their strategy underscores the need for both effective communication among system members and for the collection of varied sources of environmental data.

Application of Open-Systems Planning

Clark and Krone (1972) outline briefly four steps for applying open-systems planning in work systems. They suggest that work systems can:

> 1. Rapidly identify and map out the dynamic realities which are in their environment.
> 2. Map out how the organization represented by the members of the [system] presently acts toward and hence values those realities.
> 3. Map out how the organization wants to engage with those realities in the future [that is, to set value-goals].
> 4. Make plans to restructure the "architecture" of the [system] in order to influence the environmental realities in the valued directions (pp. 290–291).

These steps are meant to help work system members understand how they presently appreciate their environment; from this knowledge, individuals can generate alternative future environments as well as plans for implementation. Although Clark and Krone (1972) do not present evidence for the success of their strategy, they do suggest that "systems can, as a result of seeing more of and valuing differently the complicated texture of their environments, generate a considerably more varied range of action possibilities from which to choose directions and strategies" (p. 291). This method may be considered a heuristic technique for examining and making plans to act upon the environment. From our experience of using this strategy, we have a number of practical suggestions for applying open-systems planning in work systems:

1. Deal only with key parts of the environment. The tendency is to examine too much data, thus losing track of what is important for system performance and what is not. To overcome this tendency, the first step—mapping out the environment as it presently exists—should be an initial scanning which defines broad aspects or domains of the work system's environment. We usually start from a description of the primary task and then ask the question: What parts of the environment affect this task? This information is then grouped in terms of importance to the primary task; those aspects judged most significant are used for the remaining steps of the model.

2. Determine what parts of the environment are constraints and what aspects are contingencies. Since we are concerned with those parts of the

environment that are potentially modifiable, we must separate fixed constraints from changeable contingencies. This can be accomplished by examining each part of the environment identified and making an assessment as to which parts are constraints and which are contingencies. Although it is useful to identify constraints, the remainder of the analysis should focus on important and modifiable contingencies.

3. Follow the steps of the strategy in order. There is a tendency to confuse the environment as it "presently exists" with the environment as it "should be." If the method is systematically followed, step by step, the analysis will logically lead from the present environment to the future.

4. Do not evaluate alternative futures too early in the analysis. The third step of the strategy generates alternative futures or value-goals. This step attempts to answer the question. What should the environment look like if the primary task is to be performed effectively? Since we are concerned here with alternative futures, we want to generate as wide a range of alternatives as possible. Thus, our initial concern is to generate a list of alternative futures; then, from this list, we can decide which alternative is most appropriate for the work system. Premature evaluation of the alternatives is likely to constrain the range of possibilities; therefore, it is advisable to defer evaluation until a thorough list is developed.

5. Plans for restructuring the work system should account for joint optimization and developmental system design. The last step of the strategy includes restructuring the work system to influence the environment in valued directions. Since this stage involves socio-technical design, we are concerned with designing a jointly optimized work structure. Many of the structural propositions outlined previously in this chapter account for a jointly optimized work system that is also proactive toward its environment. When these propositions are combined with the method of developmental system design—a strategy providing for self-regulative work structures—the work system should have a jointly optimized structure capable of influencing its environment in desired directions.

Summary

Work systems must relate to a succession of suitable environments if they are to survive and develop. Beyond the need for boundary management, work systems must also influence their environment in favorable ways. Open-systems planning is a method whereby work systems can systematically examine their environment and take steps to actively influence it. Based on assumptions about perceiving or appreciating the environment and about the complexity of the environment itself, this method helps work system members to understand their current appreciations and ways of acting upon the environment; this knowledge is then used to formulate alternative future environments as well as ways to implement them. Much of our discussion about structures for joint optimization and about developmental system design accounts for the design of work systems that are both jointly optimized and proactive toward their environment.

SUMMARY

The purpose of this chapter is to derive management practice from the theoretical foundation for understanding socio-technical systems. Based on the concepts of directive correlation and open system theory, management practice is interpreted into a socio-technical framework. Directive correlation implies that work systems are comprised of both social and technological components. To optimize overall system performance, we must jointly optimize both components. Thus, a major function of management is to bring into existence a jointly optimized work relationship structure. This requires knowledge of the structure and process of joint optimization.

Structure is concerned with the physical arrangement of a system's components. A jointly optimized work structure accounts for both the task requirements of the technical component and the biological and social-psychological needs of workers. Several propositions for structuring jointly optimized work systems were presented. Process involves change over time. A jointly optimized work system must also account for the dynamics of technical and social change. The former is a depreciative process and the latter an appreciative process. Developmental system design is concerned with both kinds of dynamics, since it provides for an initial self-regulative work system that is able to adapt to changes in both components.

Open system theory implies that socio-technical systems must relate to a succession of suitable environments if they are to survive and to develop. In other words, work systems must match environmental conditions for effective performance. Thus, a second major function of management is to manage the system and environment relationship to assure a compatible match. This requires effective boundary conditions for environmental exchanges as well as plans for a desirable environment and ways to bring it about. The former is referred to as boundary management and the latter as open-systems planning. Boundary management requires protecting the primary task from external disruptions and regulating environmental exchanges. Beyond the need to protect the primary task and relate it to the environment, managers must also ensure that the environment itself is suitable for system performance. Open-systems planning provides work systems with a strategy for systematically examining their environment and making plans for desirable futures and ways to bring them about.

7

A STRATEGY FOR IMPLEMENTING SOCIO-TECHNICAL SYSTEMS IN ORGANIZATIONS

Based on the theory and management practice outlined in the previous chapters, the essential question is how to implement a socio-technical perspective in an actual work system. Since much of the management of work takes place in formal organizations, our starting point is an intervention strategy for introducing this approach into existing organizations. This does not preclude, however, the use of this strategy in new organizations or in other less formalized settings, such as small businesses and informal project teams. Our primary concern is to provide organizational members with a practical strategy for experimenting with socio-technical systems in their organizations.

PREMISES UNDERLYING STRATEGY

Like all new management ideas, the implementation of a socio-technical perspective involves some form of change strategy. To be useful such strategies must be as explicit as possible concerning their underlying assumptions about organizational change. The strategy offered here is based on three fundamental premises about effective socio-technical change: (1) the use of operational experiments, (2) the organizational climate, and (3) the organizational members involved.

Operational Experiments

The implementation of operational experiments in selected parts of the organization is a useful way of introducing this approach to the entire organization. By experimenting with socio-technical systems under relatively protected conditions, it is possible to gain a clearer understanding of the consequences of this approach so that decisions about its applicability for other parts of the organization can be better informed. This experimental strategy is similar to the method commonly used for agricultural innovations. The agricultural model involves field experiments carried out at various demonstration sites connected with agricultural experiment stations. New ideas are first tried out under experimental conditions, those that have some merit are then examined under normal field conditions, and the results are disseminated to interested farmers by county extension agents who serve as change agents for farming practices. A similar strategy is especially relevant for socio-technical change. Operational experiments within limited parts of an organization:

1. Allow organizational members to test different social and technological designs with a minimal amount of disruption to the overall organization.
2. Enable organizations to examine the results of various designs without the full impact of other organizational factors that usually confound the effects of existing work designs.
3. Provide decision-makers with the opportunity to see how the approach works in an actual work situation before accepting or rejecting it.
4. Give organizational members the freedom to experiment with new ways of performing work.

Organizational Climate

Organizational climate is an important determinant of effective organizational change. Since most forms of organizational change disrupt the steady state condition of the organization, the organizational climate of the work system must be conducive to such change. This is especially true for socio-technical change in which modifications in both the social and technological components frequently affect such conditions as interpersonal relationships, reward structures, work flows, status hierarchies, and job designs. If such changes are to lead to a more effective steady state, organizational members must be capable of dealing with these problems. From top management down to the shop floor, individuals who are interpersonally open, experimental, and high in trust and risk-taking behavior are better able to confront and work through the myriad of problems arising from organizational change than those who are not. Thus, an organizational climate that supports and enhances such behavior is more likely to affect successful change than one in which such behavior is thwarted.

Argyris (1972) proposes that the more a change program requires behavioral changes deviant from existing norms, personally discomforting, and entailing high degrees of variance from traditional methods, the more it is necessary to focus on interpersonal and group dynamics if such changes are to be effective. This suggests that for socio-technical change, considerable attention must be given to the interpersonal and group processes through which the change strategy is enacted. In short, we must be concerned with both the process and content of change.

Organizational Members

Since the organizational members who carry out socio-technical experimentation are often from lower organizational levels, it is important that they be actively engaged in the change process. This increases the likelihood that social and technical changes will be relevant to the specific work system, and it reduces the chances that such changes will be resisted by workers. By engaging with those individuals who perform the primary task of the work system, socio-technical change emanates from within the system. Although this implies a bottom-up approach to change, the amount of support required from all relevant organizational levels is best exemplified by an approach that gains sanction from the top and middle of the organization to engage with the bottom of the organization.

SOCIO-TECHNICAL STRATEGY

Based on the above assumptions about effective socio-technical change, a strategy for carrying out such experimentation within existing organizations has been developed. The strategy is grounded on work performed at the Tavistock Institute of Human Relations in London, England (Rice, 1958; Trist, et al., 1963; van Bienum, 1968), at the Work Research Institute, Oslo, Norway (Thorsurd, 1966), in addition to field experiments carried out by the authors in conjunction with the Organizational Behavior Department at Case Western Reserve University, Cleveland, Ohio (Cummings, in press). The strategy consists of eight sequential stages: (1) defining an experimental system, (2) sanctioning an experiment, (3) forming an action group, (4) analyzing the system, (5) generating hypotheses for redesign, (6) implementing and evaluating the hypotheses, (7) making the transition to normal operating conditions, and (8) disseminating results.

Defining an Experimental System

Defining a work system appropriate for socio-technical experimentation is a complex issue. Ideally, we are looking for a unit with the following characteristics:

1. Social and technological components clearly differentiated from other organizational units,
2. Input and output states clearly defined and easily measurable,
3. High probability for success.
4. High probability for dissemination of results, and
5. Members who are interested in experimentation.

Although it is often difficult to find an existing work system with these characteristics, it is crucial to approximate them for an initial experiment. These factors not only increase the likelihood of success, but they serve as initiating conditions for subsequent change efforts.

Differentiating a Work System

In determining if a work system is clearly differentiated from other organizational units, we are faced with the issue of bounding a system—deciding what to include in the system and what to exclude. Since this is always an arbitrary decision, it could be concluded that it is a rudimentary issue and not worthy of much attention. After all, why not take an existing organizational unit and proceed to experiment with it? If existing work systems were defined so that their social and technological components formed a relatively self-completing whole, this argument would have some merit, for it would be possible to differentiate between those relatively autonomous processes that are under a system's control and those that are relatively heteronomous and not under control. We could easily draw the distinction between system and environment.

Unfortunately, many organizational entities are not defined in terms of social and technological wholes, but rather are differentiated around part processes with components as much related to external components as to each other. One can readily imagine the problems that arise when change is attempted in one of these work units. The crux of our argument is that many work units do not meet our definition of a socio-technical system: a nonrandom distribution of social and technological components that coact in physical space-time for a specified purpose. When the purpose for defining a system is to change it, it is imperative that all components relating to each other for the performance of a primary task be included in the system definition.

Now that we have seen the importance of bounding a work system to include social and technological components functioning to perform a relatively whole task, the question arises as to how to identify such systems. A number of useful criteria exist to identify a relatively differentiated socio-technical system: time, territory, technology, and sociological and psychological attributes.

Time refers to the contemporaneous existence of components—do the components share the same time frame? Territory alludes to the physical proximity of components—do the components share the same physical space? Technology refers to components that operate the same transformation process—do the components share a common technology? Sociological attributes include individuals who share common work roles, supervision, and demographic characteristics—do the components have similar work roles, supervision, ages, religion, socio-economic status, education, and the same race and sex? Psychological attributes refer to individuals who share similar psychological states, including perceptual realities—do the components see themselves as a work system? Since socio-technical systems contain both social and technological elements, the above criteria help to account for the characteristics of each. Time, territory, and technology apply to both types of components, while sociological and psychological attributes pertain to the individuals operating the work unit.

To more clearly see how these criteria aid in the differentiation of a socio-technical system, we offer a practical example of system definition. Let us assume that we are asked to examine a sales unit of a large manufacturer of heavy-duty machinery to ascertain its appropriateness as a differentiated socio-technical system. We make a field visit to the administrative sales office and ask a few questions of the sales personnel—supervisors, salesmen, secretaries, and filing clerks. Upon first entering the sales unit, we observe the following: (1) The social components all work a standard 8.00 a.m. to 4.30 p.m. work day except for a group of filing clerks who file invoices on an evening shift for subsequent key punching. (2) The social components and some of the technological components—desks, files, typewriters, phones, desk calculators—all exist within a physical space bounded by two permanent walls and two temporary partitions. (3) The social and technological components all serve to operate a primary task of selling heavy-duty machinery. So far it looks as if we might have a well-bounded work system, with the possible exception of a group of second-shift, filing clerks.

We proceed to interview some of the personnel. Our interviews reveal that: (1) The sales force is divided into four national regions and the salesmen spend about 70 percent of their time in their respective regions. (2) Supervision consists of a department manager and four regional supervisors. (3) The salesmen and supervisors are all male with college degrees, while the secretaries and clerks are predominantly female with high school diplomas. (4) The salesmen view their jobs in terms of making face-to-face sales, and the secretaries see their work roles as providing necessary clerical services to the salesmen. (5) The filing clerks perceive their jobs as similar to the secretaries with the exception of the evening shift clerks who view their roles as preparing materials for the computer services department. (6) The salesmen spend more time relating to regional customers than they do relating to members of the sales unit, and the secretaries and day-shift clerks relate to each other more than anyone else, with the possible exception of daily phone calls from the salesmen in the field. (7) The night-shift clerks relate to each other and to the key punching operators of the computer services department. With this additional data, the work unit is not as clearly defined as we had originally envisioned.

Although the boundaries are arbitrary, the above information allows us to explore some alternatives:

1. Define the entire department as one work unit.
2. Define the entire department with the exception of the night-shift, filing clerks as one work system.
3. Define the social and technological components for each region as a system, thus generating four work units.
4. Define the salesmen and supervisors as one unit and the secretaries and clerks as another unit.
5. Define the entire department and the key punching part of the computer services department as one system.
6. Include important customers in the definition of each regional work unit.

In working through this hypothetical example, we can begin to see that the issue of where to place a work system's boundaries is far from elementary. Although our criteria—time, territory, technology, and sociological and psychological attributes—can help to differentiate a socio-technical system, the result is still ultimately arbitrary. With this in mind, we initially bound a work system and continue to raise the issue of changing this definition in light of new information.

Input and Output States

A second characteristic of an appropriate experimental system is clearly defined and easily measured input and output states. We cannot begin to ascertain the functioning of a work system if we are not able to determine the state of its inputs and outputs. Unless we can demonstrate that a particular input was converted into an output by a specified action, we cannot determine cause/effect relationships. Without knowledge of such relationships, the technical rationality of a work unit's functioning remains unknown. Moreover, if we do not know whether a system can produce a specified outcome, the question of economic rationality—whether the outcomes are obtained in the least costly manner—cannot be answered. Thus, to the extent inputs and outputs are explicitly defined and measured, our task of determining the effectiveness and efficiency of work system functioning is facilitated.

In determining the clarity and measurability of input and output states, we must often probe beyond the boundaries of a work unit. Information concerning these variables is frequently collected and stored in organizational entities designed for this purpose. Departments of accounting, quality assurance, management information systems, engineering, personnel, purchasing and stores, and planning and traffic record a variety of input and output data. Quantity, quality, size, dimensional integrity, location, weight, absenteeism, turnover, costs, and an innumerable number of other dimensions are collected for purposes of ascertaining the state of a work system. The extent to which this data is useful in assessing the effectiveness and efficiency of a work unit is a function of how well the primary task is rationally determined. When a system's primary task is rational, it knows how to produce a desired outcome. This rationality usually makes it easier to assess input and output states. Examples of rationally determined technologies are production lines, refineries, and accounting systems. On the other hand, teaching, psychotherapy, law, family counseling, and welfare services have less rationally defined technologies, and for purposes of socio-technical experimentation, their inputs and outputs are more difficult to measure.

Probability for Success

Choosing a work system with a high probability of success involves the two issues of deciding upon appropriate criteria for judgment and appraising the likelihood that a work system can demonstrate positive results. Since the criterion issue is discussed later in this chapter, let us assume that we have solved this problem and now we must determine probability of success.

One strategy is to choose a work unit that scores extremely low on the criteria, and assume that it has nowhere to go but up. Based on the premise that low scores are a valid measure of the system's present condition—that they are not extreme measures likely to regress statistically upward toward the unit's true performance, thereby producing a statistical artifact instead of actual change—this strategy has both positive and negative consequences. On the positive side, low performing systems are probably most in need of change and carrying out a successful experiment can demonstrate dramatic results. On the negative side, such units can be so beset with problems that they do not have sufficient productive slack to engage in experimentation. Also, the assumption that "there is nowhere to go but up" might be fallacious.

The converse strategy of choosing a high performing system and assuming that it is a healthy unit with a good base for improvement also has certain merits and limitations. On the plus side, the members of such work systems are often secure enough to experiment with change; they tend to seek challenges with the potential of ensuring or improving their successful image. They also usually have sufficient productive slack to allow for experimentation. On the other hand, high performing systems can be an experimental risk. Existing social relationships may be harmed, productivity can decrease, system members may resent the implication that they should improve, and the work system could be at its upper limits of performance.

Regardless of the strategy used in choosing a potentially successful experimental site, the actual choice will have both intended and unintended consequences. The reality of any experiment is the uncertainty of obtaining desired results. In this respect, there is no optimal strategy for guaranteeing outcomes; instead, we explicitly examine the merits and limitations of alternative sites and proceed accordingly.

Potential for Dissemination of Results

Since one of the purposes of a socio-technical experiment is to generate knowledge applicable to other work settings, our choice of an experimental site should take into account its potential for dissemination of results. In scientific terms, we are looking for a system with a high potential for generalization; while in terms of a strategy for diffusing innovative work principles, we are trying to find a unit that might serve as a "leading edge" for innovation.

The question of potential for generalization can never totally be resolved. No two work settings are exactly alike in respect to variables that can affect an experimental treatment. Rather than drop the issue here, there are certain considerations that may increase the probability that results in one setting may be transferred to another site. First, one can control for technology—that is, one may choose a unit with a similar primary task to other systems in the organization. Second, one can try to match work systems in terms of workers' characteristics likely to affect results. Individual differences such as seniority, sex, education, and alienation from middle-class norms are prime targets. One can also attempt to control for environmental variables. Plant location, unemployment rate, market condition, and plant climate are relevant variables. Finally, it is possible to control for leadership style, since the way in which a manager behaves toward workers may have an impact on results. The best advice for addressing this issue is to be aware of variables that may potentially affect experimental treatments and try to choose a unit that is similar on these dimensions to other units where one might want to transfer the results.

Judging the extent to which a work system is a leading edge for innovation is important for expanding a socio-technical approach to a wider organizational context. In choosing an initial experimental site, we are trying to initiate conditions for wider systematic change. Current theory in diffusion of innovations is limited primarily to the marketing of new products and services—research on diffusion of agricultural and medical innovations is a prime example. A significant conclusion of these studies is that individuals who are most likely to adopt new practices are those who most actively seek new knowledge about innovations, and if this knowledge is to have a multiplier effect, these individuals must also be influential.

In relation to a "leading edge" system for socio-technical diffusion, the work unit must seek new ways of doing things, and it must have a key position in the influence structure of the organization. Determining the amount of influence a work unit possesses requires knowledge of the influence network of an organization. One must know the formal structure of influence, exemplified in an organization chart, as well as the informal networks along which unofficial sources of change flow.

Interest in Experimentation

A final characteristic of an ideal experimental system is that the members are interested in taking part in socio-technical experimentation. Since we experiment with people rather than on them, we strongly suggest that workers and managers be given the opportunity to make their own choices concerning experimentation. Because of the ambiguous nature of socio-technical experiments, it is difficult to present an accurate portrayal of an experiment in advance. Instead, we attempt to provide pertinent information concerning our approach and reasons why we are considering a particular unit for experimentation. At best we hope to arrive at a tentative agreement among workers and management to actively engage in the analysis and redesign of their work system. Depending upon the amount of interest shown, a final choice is made with the agreement that the issue of interest will be raised at periodic intervals. In this respect, the possibility of termination of an experiment is left open for further negotiation.

Sanctioning an Experiment

Once an experimental site is chosen, terms for carrying out the experiment must be specified and sanctioned by workers and management. The primary purpose for sanctioning an experiment is to provide the protection necessary for experimentation. Normal organizational demands and constraints that impinge on the work unit must be temporarily suspended if new designs are to be tested under experimental conditions; workers must be afforded wage and job security if they are to feel free to experiment without fear of losing money or a job, and contractual arrangements, such as job classifications and wage scales, must be open to potential change if hypotheses for redesign are not to be limited severely. In effect, socio-technical experiments require the same if not more protection than the start-up of a new work system.

The primary means of providing protection is by official sanction from the highest organizational levels directly affecting the experimental system. Sanctioning from such levels requires the active concern of workers (including union officials if this is the case) and managers (both line managers and managers of other functional units directly relating to the experimental system). Since the sanctioning body provides a protective umbrella, its membership should include individuals with control over factors that may affect the work system. Without official sanctioning by workers and managers alike, the experimental system would be subject to the full range of forces that affect a normal operating unit, thus reducing the chances for successful experimentation.

The process of obtaining sanction necessary for socio-technical experimentation involves three related issues: (1) types of protection, (2) sanctioning body, and (3) sanctioning rules.

Types of Protection

Herbst (1956) identifies three forms of socio-technical protection: (1) conceptual, (2) experimental, and (3) operational.

Conceptual protection is required during the analysis and design stages of an experiment. This form of protection should allow members complete freedom to explore all dimensions of their work unit and to devise new ways of dealing with problems. A consequence of not providing complete conceptual protection is that members may feel constrained in exploring the full range of problems and possibilities for redesign.

Experimental protection is required during the implementation and evaluation phases. This type of protection should permit the unit substantial exemption from normal operating demands so that hypotheses for redesign can be examined and modified without undue requests for productivity. Failure to provide experimental protection often results in premature closure of work-system changes because of the anxiety aroused by demands for productivity.

Operational protection is required during the transition phase from experimental to normal operating conditions. Workers must be given sufficient time to bring new modifications up to normal operating performance. Since social change often lags technological change, sufficient time should be allowed for a gradual break-in period. This will vary according to the magnitude of the change and the adaptive capacity of the work force.

Sanctioning Body

Because of the importance of a protective umbrella for socio-technical experimentation, appropriate membership on the sanctioning body is crucial if the sanctioning group is to have control over those factors that can adversely affect the experimental system. To ensure such control, one must determine the highest implicated level that can directly affect the unit, and then elicit the support of members from this level. Prime targets for membership include operational managers, managers who supervise other functional units directly relating to the experimental system, and official union representatives if the work force is formally organized. In those case where workers are not organized, a method for determining worker participation on a sanctioning body must be devised. Although we have little control over the level chosen, we urge workers to sanction at the highest level that can have an impact on the experiment. In many instances, this may include national union officials. A rule of thumb for membership on the sanctioning body is to ascertain the highest organizational and union levels with a direct impact on the experimental system and include anyone from these levels who might help or hinder the experiment. We feel that it is better to identify and work with such individuals during the sanctioning process than after the start of the experiment.

Sanctioning Rules

In providing experimental protection, the sanctioning body must decide upon specific rules for guiding the experiment: the length of time and amount of each type of protection, specific guarantees as to wage and job security, plans for those workers who do not want to experiment or who cannot adapt to new designs, and the extent to which organizational policy and contractual obligations must be followed. Depending upon the circumstances, provisions for union representation, seniority, bidding order, and other formal regulation must be managed. In lying out the "rules of the game," workers and management agree on the boundaries of the experiment in sanctioning these tenets, they are saying that "within these bounds we are protecting your freedom to experiment." Without a clearly specified and officially sanctioned set of rules, socio-technical experimentation would not have the necessary organizational support needed for meaningful change.

Forming an Action Group

Socio-technical experiments are often carried out under the direction of a small group composed of workers and first-line managers of the experimental unit. The purpose of the action group is to perform the analysis and to generate hypotheses for redesign. This group may also supervise the testing or evaluation phase of the experiment. In either case, the action group serves as the primary link between the experimental system and the sanctioning body. In performing its function, the action group collects and analyzes data, proposes new designs, and evaluates results. It also keeps the sanctioning body informed of experimental progress and serves as a sounding board for workers and managers involved in the experiment.

Action groups are utilized when the size of the experimental system is too large for all members to directly supervise the experiment. We recommend such groups when the size of the unit exceeds eight to ten workers. Since experimental activities are often perceived as complex and difficult to understand, help is sometimes needed from appropriate resource people. Although the extent of such aid will vary according to the needs of the experimenters, we believe that the basic process of socio-technical experimentation can be learned and carried out by most organizational members within a relatively short period of time. To expedite the experimental process, the action group should be limited in size from three to six members—first-level supervisors and a few interested and committed workers. During the analysis and hypotheses generation phases, some if not all members of the action group should devote their full time to carrying out experimental activities. This provides the impetus necessary for starting an experiment and serves as a positive sign of organizational support.

Finally, since the action group represents the experimental system, all efforts should be made to keep workers informed of its activities. Without full cooperation and support from all members of the work unit, the data base for analysis and redesign is limited severely.

Analyzing the System

The analysis of a socio-technical system requires a framework for collecting and organizing data. One must not only know what information to gather, but how to put it into an understandable form. Based on extensive work carried out at the Tavistock Institute of Human Relations, two analytical models have been developed for this purpose: (1) a process or production system model and (2) a model for those work systems where no continuous process exists—e.g., service units.Because of the detail of these frameworks, each is presented in the Appendix. Our aim here is to outline briefly the two models and to discuss key issues for performing an analysis.

Production System Model

The process or production system framework contains six steps:

1. *Initial scanning*—The main characteristics of the work unit and its environment are identified in order to determine where the primary emphasis of the analysis should be placed.
2. *Identification of unit operations*—The principal stages in the production process are located and each identifiable segment is viewed as a unit operation with its own set of systemic properties.
3. *Identification of key process variances and their interrelationships*—The purpose of this step is to identify variances that arise from the production process or from the nature of the raw material. A variance refers to any deviation from some standard or specification, and a variance is termed "key" if it significantly effects quantity or quality of production, or operating or social costs.
4. *Analysis of the social system*—The main characteristics of the social system are examined, and a table of variance control is constructed to determine the extent to which key variances are controlled by the social system. Ancilliary activities, spatio-temporal relationships, job mobility, and the payment system are also examined, and existing work roles are tested against a list of basic psychological needs.
5. *Workers' perceptions of their roles*—An inquiry is made about how workers perceive their roles with special reference to the degree to which the basic psychological needs are met.
6. *Environmental analysis*—we then turn to the environment of the work system and examine how other units affect its functioning. Specifically, we look at relations with the various support systems in addition to the supply and user systems. Our purpose is not to examine these units as socio-technical systems, but to discover how they affect the experimental system. Finally, the work system is considered in the context of the larger organization with particular attention to the effects of developmental plans and general policies.

Service System Model

The service system model also contains six steps (those steps similar to the steps in the previous framework are not elaborated upon):

1. *Initial scanning.*
2. *Objectives of the system*—The aim of this step is to clearly

identify system objectives so that activities can be judged in terms of appropriateness for meeting such goals. Against this background, we can begin to hypothesize necessary responsibilities, authorities, communication links, and methods and procedures.

3. *Role analysis*—The objectives of each work role are ascertained, and this information is related to the system's objectives previously identified. This matching enables one to discover incongruities between the goals of the work role and the objectives of the experimental unit.
4. *Grouping of roles*—Role interactions are identified in the context of work flow, and hypotheses are generated for clustering these roles in respect to their temporal, geographical, and status dimensions.
5. *Worker's perceptions of their roles.*
6. *Environmental analysis.*

Performing an Analysis

The application of an analytical model may be facilitated if certain guidelines are followed. First, the experimental system should be examined in its current state. This provides a base line for generating hypotheses for change and avoids confusing "what is" with "what was" or "what could be." Second, data should be collected from a variety of sources. Workers, managers, and appropriate specialists offer different views which can be useful for determining the structure and functioning of the work system. Third, analysts often get into too much detail and, as a consequence, lose sight of the main problems of the system. A concerted effort should be made to list only key information and to avoid over elaboration of data. Paying particular attention to the initial scanning may help to eliminate unnecessary information. Fourth, analytical data should be cross-validated for accuracy by allowing the data sources to view and discuss preliminary conclusions. In this respect, the analytical process should always be open to critical dialogue. Finally, hypotheses for redesign are often generated during the early stages of the analysis. These should be listed and carried forward into the next phase of the experiment.

Generating Hypotheses for Redesign

The primary outcome of the analytical phase of a socio-technical experiment is to generate hypotheses for job or work system redesign. Although a variety of possible changes may be made, we are particularly concerned with hypotheses aimed at joint optimization of the social and technological components and at a better matching between the work system and its environment.

Creating redesign proposals involves a process of collecting and analyzing information and recombining the data into new configurations to produce alternative designs. In short, one creates the "new" from the "existing." Since such syntheses evolve from knowledge of a problem, the analytical phase of the experiment must provide a clear statement of the problems of the work system. Once the designer is aware of the goals—social and economic—of the system, the difficulties to be overcome, the resources available, and the constraints which determine an acceptable solution, he can begin to generate redesign proposals. Although Chapter 6 listed structural propositions for jointly optimized designs, the content of specific designs must be relevant to the particular setting of the experimental unit. Thus, it is not possible to transplant carbon copies of one particular design to other work systems, since each redesign must be tailored to the needs of the experimental system.

Since socio-technical design is carried out by individuals closely associated with the experimental unit, the process of generating hypotheses is often perceived as a novel and difficult project. A major barrier to the generation of alternative designs is premature evaluation of individuals' ideas. Such evaluation frequently results in withdrawal from the design process. For example, Joe proposes a possible redesign, and Harry immediately evaluates Joe's proposal by saying, "It won't work." Joe tentatively agrees and then withdraws from the creative process. This withdrawal problem may be overcome if, during the early stages of proposal generation, members of the action group limit their activities solely to formulating ideas for change. Since one hypothesis may generate others, members should be encouraged to share their ideas—no matter how outrageous—and to build on the propositions of others postponing all evaluation.

A problem often arising in this stage of an experiment is workers' concern that proposals for change imply that something is wrong with their present performance. Although this issue is usually not talked about and is often not consciously understood, it is frequently manifested by a reluctance to generate new ideas. Though we can offer no immediate solution to this problem, an awareness of its existence may help to resolve it.

Implementing and Evaluating the Hypotheses

Now that we have generated hypotheses for redesign, we must proceed to test or evaluate them under experimental conditions. This phase of a socio-technical experiment involves four stages:

1. Reducing the list of possible hypotheses to a manageable set for experimental testing,
2. Devising an action program for the testing period,
3. Implementing the action program, and
4. Evaluating results.

Choosing a Manageable Set of Hypotheses

Since a variety of alternative hypotheses is often available for consideration, it is necessary to reduce this list to a more manageable set for experimental testing. Deciding which alternatives are most promising involves a judgment as to their viability against appropriate criteria. Although the actual criteria will vary from one context to another, they must relate to the experimental system's objectives. In other words, they must pertain to production goals—quantity, quality, operating costs—or to social objectives—meeting psychological needs, reducing stress, accidents, absenteeism, turnover. Since some proposals deal exclusively with one kind of objective while others involve a mixture of both, the hypotheses must be identified specifically as to their intended effects. Once these effects are identified, an assessment as to their likelihood for achievement must be made.

Given that one can ascertain the probability for a proposal producing results—at least at the subjective level—one must then deal with the issues of cost. Can the redesign achieve a desired outcome without undue social and economic costs? In determining costs, it is necessary to project both short- and long-term consequences, since the social and economic effects of any proposal may take considerable time to accrue. Deciding to reduce preventive maintenance, for example, may lower production costs in the short-run, but the long-run effects of reduced machine life may far outweigh the initial benefits. One of the primary difficulties in calculating costs is that many proposals produce outcomes which are multi-valued because individuals often place different values on the same result. One hypothesis may reduce work hazards at the expense of productivity. Depending upon one's position—social responsibility versus economic rationality—the proposal will be evaluated differently. While there is no optimal way to resolve a multi-valued choice, the extent to which redesign proposals satisfy both social and economic interests can help.

Although the primary criteria for deciding upon the utility of a redesign proposal are instrumental and economic—can the proposal produce the desired outcome, and is the outcome obtained in the least costly manner—three additional criteria must be considered: (1) potential spill-over effects, (2) meaningfulness of the redesign, and (3) elegantness of the redesign.

Since all socio-technical systems are embedded in an environment composed of other work systems, it is necessary to examine the potential spill-over effects of an organizational change. In determining them we must examine both intended and unintended consequences. Depending upon the degree to which the experimental unit is related to other parts of the organization, spill-over effects may extend to diverse parts of the organization. Although it is impossible to project fully the wider consequences of a redesign, a concerted effort should be made to include such considerations.

In deciding among alternative proposals, it is useful to gauge the extent to which system members perceive the redesign as meaningful. Since workers possess intimate knowledge of the social and technological aspects of their work system, they can serve as barometers for a design's meaningfulness. Their perceptions can also provide some indication of potential implementation problems.

A final criterion for evaluating hypotheses is the extent to which the design is elegant. By elegant we mean a design that is simple or precise in its solution to a complex problem. Given two proposals relatively equal on all other criteria, we prefer the one that is more simple or elegant. Not only do we believe that such designs are more aesthetic and intelligent, but they are more easily understood, implemented, and evaluated.

Devising an Action Program

Once a decision is made concerning the redesign to be tested, a plan for implementing it under experimental conditions must be devised. An action program serves as a framework for introducing the changes in a systematic manner so that their impact can be evaluated properly. Since the content of this plan is crucial, it should include specific reference to the following:

1. A detailed listing of the proposed changes,
2. A timetable for the introduction of each change,
3. A prospectus of the conditions for experimental protection,
4. An inventory of the services, tools, and materials needed for experimentation,
5. A timetable for evaluative activities (instrumentation, analysis, and feedback),
6. A listing of special rules or procedures that apply to the experimental period (job postings, vacations, sick leave, etc.),
7. A determination of training needs and a program for meeting these needs, and
8. An account of supervisory functions and responsibilities during the experimental period.

Inasmuch as these details go beyond the boundaries of the experimental system, the sanctioning body should take a decisive role in the creation of an action program. Without their active engagement and support, the protection necessary for experimentation would be difficult, if not impossible to attain.

Implementing the Action Program

Implementing a work system change requires an inordinate amount of patience and energy. Members of the experimental unit must not only learn the new, but they must also learn to forget parts of the old. Although the difficulty of this process varies according to the characteristics of the target system and of the change proposal, there are at least four relevant issues.

1. *Timing the change* is critical for discontinuing the existing and starting the new. One should take advantage of naturally occurring disjunctions—the start of accounting periods, the end of vacations, the first of the month, the end of a preventative maintenance shut-down, the change to a new product, etc. These breaking points serve as perceptual boundaries for workers, thereby enabling them to experience the existing cycle of events as coming to a natural completion before embarking on a new endeavor.

2. *Focusing the resources* necessary for getting a change underway is important during the early stages of implementation. Providing the required tools, equipment, services, and other resources is essential for eliciting the synergy needed to initiate change. There is nothing more damaging to a socio-technical redesign than starting without the necessary resources.

3. In carrying out socio-technical change, the *speed of social change*

often lags that of technological change. As we have already discussed, the technological system can be implemented in one step, while the social system progresses through a series of intermediate stages as it develops towards maturity. A primary reason for this longer developmental process is that members of the social system must perceptually reorient themselves. Developing new role relationships and learning how to perform different tasks often require a good deal of time. During the experimental stage, productivity may fall below the normal operating level as workers modify their behavior while experimenting with novel conditions. Although we have no adequate way of judging how long this will take, the socio-technical experimenter should be forewarned and consider it a normal part of the experimental process.

4. Modifying a work system frequently results in *resistance to change.* Since existing work roles are invested with a high degree of status and emotionality, we can expect that anything disruptive will be resisted. Instead of interpreting this as resistance to change, we can view it as an active seeking toward positive affirmation. To the degree a redesign negates existing sources of affirmation, workers will actively seek to preserve those roles that provide them with positive feelings. Thus, new designs should provide workers with increased opportunities for status and positive affirmation if this source of energy is to contribute to effective change.

Evaluating Results

All socio-technical redesigns are hypotheses that must be tested for their effectiveness in the particular organizational setting in which they are employed. In evaluating a redesign, it is necessary to deal with three issues:

1. The criterion problem—on what grounds are we to measure the effectiveness of possible designs?
2. The instrumentation problem—can we validly and reliably measure the criteria?
3. The research design problem—can we design an experiment to draw valid inferences about the effects of the experimental treatment?

The criterion problem concerns the choice of measures of effectiveness. The primary concern of a design proposal is to help a work system meet its social and economic objectives. Inasmuch as the evaluation of any redesign implies some form of hypothesis—either implicitly or explicitly formulated—which states that a specific design is expected to result in a particular outcome, we must relate this outcome to the work system's goals. In short, criteria of effectiveness must reflect what we are trying to achieve—goal accomplishment.

In deriving criteria, we are often confronted with the problem of choosing which aspects of the work system should serve as appropriate measures of performance. Since many objectives are multi-dimensional, we must frequently include several criteria for the same goal. Thus, reducing overhead costs may involve a wide range of related costs—supervision, inspection, maintenance, etc.; to adequately measure this objective, we would have to include these related variables. While many of the economic objectives can be measured directly, the social goals frequently involve indirect or unobtrusive measures. Psychological stress can relate to absenteeism, turnover, accidents, grievances, and the like. Therefore, in choosing measures of effectiveness, one should attempt to determine those variables—either direct or indirect—that bear on the system's objectives and include all or a representative sample as criteria. Webb, Campbell, Schwartz, and Sechrest (1966) present a thorough discussion about indirect measures of social behavior.

The instrumentation problem involves measuring criteria of effectiveness. Once we have chosen appropriate criteria, we must then face the question: Can we validly and reliably measure them? In regard to instrumentation, validity refers to the ability of an instrument to measure what it purports to measure, while reliability refers to the consistency of these measures over time or across tests. The use of valid and reliable data collection techniques is critical in evaluating redesign experiments. If an instrument is not measuring what it claims to measure or if it is unstable in obtaining data, one has no assurance of what a redesign actually accomplished.

For many of the economic criteria, data collection appears fairly rudimentary. One collects pertinent information from existing organizational records with the assumption that it is valid and reliable. Although we often take such assumptions for granted, much so-called hard data are, in fact, subject to a wide range of viscissitudes. One example is that machine utilization is often recorded by machine operators who vary in reliability across individuals and across different time periods. Another example is the variance that arises in different data collection categories. For instance, it has been shown that suicide rates and crime rates vary according to the way they are classified—one way to reduce crime is to change your definition of crime.

The collection of social data is often a more difficult task than obtaining economic information. Standard techniques for measuring psychological need achievement, stress, alienation, involvement, satisfaction, learning, human development, and the like are often difficult to apply and to interpret. If one wishes to devise his own instrument for this purpose, he is again faced with issues of validity and reliability. Although we do not mean to discourage a socio-technical experimenter, we do feel that the importance of measuring validly and reliably the criteria of effectiveness is imperative. Lake, Miles, and Earle (1973) provide a valuable review of instruments for collecting social data.

The research design problem concerns designing a socio-technical experiment to determine the effects of a particular redesign. Again, we are faced with the issue of validity, but here it refers to the question: Did the experimental redesign produce the observed effects? We refer to this as "internal validity" and to the problem of potential for generalization as "external validity." Since the potential for generalization is irrelevant without internal validity, our discussion focuses on this latter problem.

The primary purpose of this stage of experimentation is to examine the effects of possible redesigns. In doing this, we must test each redesign as a hypothesis for change. In determining the effectiveness of a hypothesis, we are, in fact, rejecting alternative hypotheses. That is, in order to claim that a redesign had an effect on our experimental unit, we must show that extraneous variables—in the form of rival hypotheses—did not produce the observed results. Campbell and Stanley (1966) identify eight types of extraneous variables that, if not controlled in the experimental design, can produce confounding results. These include:

1. *History*, the specific events occurring between the first and second measurement in addition to the experimental variable.
2. *Maturation*, processes within the respondents operating as a function of the passage of time per se (not specific to the particular events), including growing older, hungrier, more tired, and the like.
3. *Testing*, the effects of taking a test upon the scores of a second test.
4. *Instrumentation*, changes in the calibration of a measuring instrument or changes in the observers or scorers may produce changes in the obtained measurements.
5. *Statistical regression*, operating where groups have been selected on the basis of their extreme scores.
6. *Selection*, biases resulting in differential selection of respondents for the comparison groups.
7. *Experimental mortality*, differential loss of respondents from the comparison groups.
8. *Selection-maturation interaction, etc.*, which in certain of the multiple-group, quasi-experimental designs is confounded with (i.e., might be mistaken for) the effect of the experimental variable.

Campbell and Stanley show that these variables can best be controlled in a true experimental design in which the researcher has full control over the scheduling of the experimental treatment—"the 'when' and 'to whom' of exposure and the ability to randomize exposures" (p. 34). Since few, if any, socio-technical experiments fall into this category, we must turn to a weaker yet valuable alternative—a quasi-experimental design. Quasi-experimental designs are appropriate when the experimenter has some control over the scheduling of data collection—the when and to whom of measurement. Although they are not as strong as true experiments, quasi-experiments allow for the control of some of the external variables that are plausible threats to the validity of an experiment. Thus, they are useful whenever better designs are not feasible. Srivastva, Salipante, Cummings, Notz, Bigelow, and Waters (1975) provide an excellent account of the use of quasi-experimental designs in work experiments.

The importance of utilizing an appropriate research design in socio-technical experimentation cannot be overstated. Given the amount of resources required for a typical experiment, we believe that the experiment should be designed so that results can be evaluated adequately. Experimental design issues should be among the first problems resolved before any experimentation gets under way. This should allow full advantage to be taken over the scheduling of data collection, and it should alert the experimenter to possible research problems. In lieu of writing a text on socio-technical experimental design, we strongly urge work system experimenters to read thoroughly Campbell and Stanley (1966) and Cummings, Glen, and Molloy (1976) before starting an experiment.

Making the Transition to a Normal Operating System

The transition from experimental testing to normal operating conditions involves a limited amount of experimental protection. Since the length of this handing-over process depends on the magnitude and type of change—larger deviations from the traditional and social change taking longer—it is necessary to take these variables into account when determining the schedule for the gradual reduction of protection. Inasmuch as some experimentation takes place during the transition period, the boundary between these two phases is difficult to determine. Therefore, much of the previous discussion concerning social and technological change (presented in the section on Implementing the Action Program) is applicable during the transition period.

Since the transition to a normal operating condition takes place only after the unit has been operating successfully, it is advisable to continue evaluative activities for some time after the experimental stage. This provides information as to the ability of the unit to remain in its new steady state, and it furnishes a data base for subsequent change. Continued evaluation may be thought of as an on-going feedback process that is built into the work system. In effect, the evaluative criteria are used as indicators of socio-technical functioning. To the extent that these indicators are periodically employed and the information utilized by members of the work system, the experimental unit has an on-going critique of its operations.

Disseminating Results

Once the experimental system is operating under normal working conditions, the process of disseminating results to other organizational units begins. Although experience in diffusion of socio-technical results is limited, Walton (1975) presents a thorough analysis of eight work experiments' dissemination efforts. Since his study provides useful insights into the diffusion process, let us examine his research in detail.

The eight work experiments follow an orientation similar to the one presented here. Since they include a variety of technologies, work settings, and national cultures, knowledge of their dissemination experience appears to have high potential for generalization. Walton (1975) identifies ten reasons why the experiments' dissemination was not as rapid or extensive as expected:

1. Some experiments failed to sustain initial success, thus eroding company-wide support for further change projects.
2. Several of the pilot projects were poor models for change

because they lacked either visibility or credibility. Here, for instance, the failure to choose a widely applicable experimental system caused outsiders to question its applicability to other situations.

3. The management of certain units tended to be too abstract or too specific in disseminating information; rather than communicate clearly the aims of such studies, they often focused on particular techniques.
4. The concepts or metaphors used to describe the changes were frequently unrealistic for other units in the company. In one case, the concept of autonomous groups was not feasible for other departments in the organization.
5. Several experiments failed to provide adequate knowledge about how to implement their approach in other parts of the company.
6. Top management often reduced their support for long-term-experimentation as environmental and organizational changes diminished either the need or the opportunity to experiment.
7. Some unions presented obstacles to dissemination as jurisdictional disputes and historical patterns of bargaining thwarted the wider application of work restructuring.
8. Bureaucratic barriers to diffusion of results included such forces as vested interests, rigid company policy, and insufficient power to change related practices.
9. Many of the work changes threatened the existing work roles of other personnel, especially first-line supervision.
10. Several of the more comprehensive experiments tended to be self-limiting. For instance, experimental participants often felt special and saw their unit as superior to other work systems; while this helped the experiment, it left outsiders with the impression that the culture created by the experiment was unique and not applicable to other contexts.

Given these situational factors as barriers to the dissimination process, Walton (1975) reviews relevant literature on diffusion of innovations and presents a number of attributes appearing to influence their adoption rate. First, innovations with a relative advantage over existing practices are more likely to be adopted than those without. Second, diffusion is enhanced if the innovation can be explained easily and if its results are readily separable from other confounding factors. Third, dissemination is facilitated if the new practice is compatible with existing norms, values, and structures. Fourth, innovations that affect fewer aspects of the system are more rapidly diffused than more pervasive ones. Fifth, changes that can be adopted experimentally and reversed without serious consequence are more readily adopted than those which cannot be reversed easily. Finally, innovations requiring few channels of approval are dissiminated more rapidly than those demanding numerous sources of acceptance. In regard to the eight work experiments, Walton (1975) concludes that they have many attributes making their dissemination inherently slow:

> Even if they offer relative advantages over existing work structures, their character and results are not highly communicable; they are not congruent with existing norms and values; their potential effect in a given work situation is pervasive rather than fractional; they are not readily reversed without incurring social costs; and too many affected parties serve as gatekeepers for the effective implementation of the innovations (p. 21).

Walton's analysis is useful for assessing the dissemination potential of socio-technical experiments. Furthermore, his study uncovers a number of problems that, if left unattended, may limit the wider acceptance of work restructuring projects. Careful planning is required to ensure that the experimental unit: continues to show good results, is sufficiently visible and clearly convincing, is accountable in the organization, provides know-how for implementation, and is supported by powerful groups in the organization. Although the experimental context will determine, in large part, how these conditions are met, it seems preferable to address them before long-term experimentation is considered.

SUMMARY

The implementation of a socio-technical perspective requires a change strategy for introducing this approach into existing organizations. The strategy presented here is based on three premises: (1) the use of operational experiments in limited parts of the organization, (2) an organizational climate responsive to both the process and content of change, and (3) sanction from the top and middle of the organization to engage with the bottom of the organization. Based on these assumptions about effective organizational change, an eight-step strategy for socio-technical experimentation has been developed:

1. Defining an Experimental System—a unit with clearly differentiated components, definable inputs and outputs, a high probability of success and dissemination of results, and members who are interested in experimentation.
2. Sanctioning an Experiment—providing experimental protection from the highest implicated levels in the organization.
3. Forming an Action Group—identifying those organizational members who carry out the experimental activities.
4. Analyzing the System—collecting and organizing data to determine the system's problems and redesign needs.
5. Generating Hypotheses for Redesign—recombining the analytical data into alternative designs for system improvement.
6. Implementing and Evaluating the Hypotheses—deciding upon an appropriate redesign, devising an action program for

implementation, implementing the action program, and evaluating results.

7. Making the Transition to a Normal Operating System—reducing experimental protection and providing for an ongoing critique of the system.
8. Disseminating Results—expanding the approach to other organizational units.

Given this strategy for socio-technical change, Chapters 8 and 9 examine its use in two work experiments.

8

THE ESTIMATING AND DIE ENGINEERING EXPERIMENT: A CASE STUDY OF WHITE COLLAR DEPARTMENTAL REDESIGN

The previous chapter presented a strategy for applying a socio-technical approach in organizations. Let us examine this strategy in an actual work setting: a socio-technical experiment involving white collar workers in an aluminum forging plant. Specifically, the study concerns two departments—estimating and die engineering—that were combined into a single unit for purposes of experimentation. The various phases of the project took approximately one and one-half years to complete; each stage is presented in sequential order, though the actual experimental events were not as orderly as they may appear. This is a pertinent point, for socio-technical experimentation is not a determinate process precisely following the framework presented in the previous chapter. Rather, it proceeds from sound theory and practice to considerable innovation on site as new information is gathered and emergent conditions are encountered. Here, we are concerned with portraying an experiment as a rational process carried out under conditions of uncertainty. Relevant theory and practice from the previous chapters are discussed in the context of the experiment to provide a better understanding of the socio-technical approach.

BACKGROUND

The X Company is engaged in aluminum and titanium forging. Forging consists of placing a piece of metal between the jaws of a press and shaping it into a particular form by the use of mechanical or hydraulic pressure. The shape of a forging is determined by the design of a die that serves to mold the metal into a particular configuration. The estimating (EST) and die engineering (DE) departments are an integral part of the X Company, since they provide valuable support services. The primary task of the estimating department is to estimate the cost of a forging for a potential customer. The output of the EST department, in the form of an estimate, is then used by the sales department as part of a bid package to obtain new business. In this sense, the EST department's duties are primarily pre-sale. The die engineering department's primary task is to design the dies used to shape a customer's forging. The output of the DE department, in the form of a die blueprint, is used by the die shop to sink a die into a piece of steel. Thus, the DE department's duties are primarily post-sale and pre-production. Both departments also provide other services.

The socio-technical experiment was undertaken as part of a system-wide, organization development program carried out at X Company in conjunction with the Organizational Behavior Department at Case Western Reserve University. The latter group served as outside consultants during this effort. The primary impetus for socio-technical experimentation came from the two top levels of management; they perceived a need to restructure one or both departments to obtain a better match between the requirements of the organization, the tasks of estimating and die designing, and the psychological characteristics of the workers. The following reasons were stated for the experiment:

1. An unacceptable ratio of estimates to orders received of the approximate magnitude of 1:10.
2. An unacceptable discrepancy in the estimated cost of a forging and its actual cost.
3. A feeling that both departments were seen as dead-end places to work.
4. A lack of new and creative approaches to forging and designing dies.
5. A lack of integration in getting new products started and delivered satisfactorily to customers.
6. A suspicion that many of the jobs were unrewarding and poorly designed.
7. A low regard of the estimating and die engineering departments by other departments dependent upon them for services.

The steps followed in using a socio-technical approach to redesign the estimating and die engineering departments include: (1) identifying the socio-technical system, (2) sanctioning the experiment, (3) forming an action group, (4) analyzing the system, (5) generating hypotheses for redesign, (6) implementing the redesign, and (7) evaluating the experiment.

IDENTIFYING THE SOCIO-TECHNICAL SYSTEM

The decision to experiment with the EST and DE departments was based on a number of subjective judgments. First, management felt that the problems identified with these units required some form of departmental redesign. A socio-technical strategy seemed promising because these problems were tied to both social and technological components. Second, the EST and DE workers indicated a preliminary interest in trying something new. It was also anticipated that the results of the experiment would be applicable to other white collar departments in the X Company—e.g., plant engineering, planning and traffic, and accounting. Given these judgments, the task of bounding the units for socio-technical experimentation was undertaken. Here, two problems were encountered: the choice of an appropriate system boundary and the relative lack of clearly defined and easily measurable inputs and outputs.

The choice of an appropriate system boundary rests on the assumption that a clearly differentiated unit provides the task coherence required for socio-technical experimentation. The consequences of poor judgment at this stage of experimentation are great, for an inadequate boundary definition may place the analysis and redesign problem outside the boundary of the work system. The boundary issue was a major problem in this study. Under the existing structure of the X Company, the EST and DE departments reported to separate functional units; the former was located under the production department, while the latter reported to the development department.

Given this organizational arrangement, two experimental boundaries were possible: either each department could be taken separately, or they could be combined. To decide upon an appropriate system for experimentation, interviews were conducted with the supervisors of each unit. These interviews focused on the work flow and role structure of each system as well as task interactions within and across departmental boundaries. Using the boundary criteria of time, territory, technology, and social and psychological attributes, the interview data revealed that on certain dimensions, the two departments were separate systems, while on other factors they could more appropriately be viewed as one department. On the side of the separate-department hypothesis, each unit carried out a different primary task, and members from each department perceived themselves as belonging to a different work system. On the side of the combined-department proposition, the workers shared a similar territory, separated only by file cabinets and desks, they worked similar hours, 8 a.m. to 5 p.m., many of their tasks were interrelated so that individuals interacted frequently with each other during the work day, and several workers had similar sociological attributes in terms of age, sex, education, and socio-economic status.

Since the interview data did not yield a clear answer to the boundary problem, the final decision was more arbitrary than expected. Because of the task interrelationships between the departments, it was felt that separate experiments undertaken in each unit would interact to create potential problems between the departments. Thus, it was decided to combine EST and DE into one work system, under the direction of the development department. Given the importance of having a well-bounded experimental system, it was anticipated that this decision would provide a clearly differentiated unit; however, since the choice of a combined department was based on ambiguous data, there were some reservations.

Once the boundary problem was resolved, the issue of clearly defined and easily measurable inputs and outputs was confronted. Here, it is important to have clear measures, for without them the rational functioning of the experimental system is difficult to determine. Inherent in many white collar jobs, such as EST and DE, are ambiguous measures of inputs and outputs. For instance, the number of inputs and outputs may be counted easily, but their quality and utility are difficult to determine. This makes it difficult to relate a worker's behavior to the results of his performance; without this connection, the feedback necessary for goal-directed activity is relatively absent. Since inputs into the EST and DE departments are highly variable (often constituting new forms of forging work) and because outputs contribute to decisions made in other units of the company, it is extremely difficult to attribute success or failure to a worker's behavior. Because of this lack of clearly defined inputs and outputs, it was decided to approach this problem as part of the analytical phase of the experiment. It was anticipated that the analysis would provide a scheme for measuring adequately the inputs and outputs of the EST/DE unit.

In summary, the decision to experiment with the EST and DE departments was based on a number of subjectively assessed criteria. Given an initial exploration of the units, it was felt that: (1) task interrelationships between the departments necessitated a combined experimental system, (2) a lack of clear inputs and outputs, though not ideal, could be overcome through proper analysis, and (3) the probability of success, chances for dissemination of results, and workers' interest in experimentation were adequate. Given these judgments, it was decided to proceed with sanctioning the experiment.

SANCTIONING THE EXPERIMENT

Sanctioning the EST/DE experiment involved a series of meetings in which management decided on the general parameters for the experiment and then conveyed these rules to the workers for their discussion and approval. During the early phases of this process, managers met alone to work out wage and job security issues. These meetings included three levels of management: the works manager, the head of the development department, and the supervisors of the EST and DE departments. The following rules were agreed upon by management and presented to the workers:

1. No one would lose his job as a result of the experiment. If workers redesigned their jobs so that fewer individuals were needed, comparable jobs would be made available in other parts of the company.
2. Present wage levels, including standard raises, would be maintained.
3. All work connected with the experiment would take place on company time.
4. The experiment would be supervised by an action group composed of first level management and selected workers.
5. Evaluation of the progress of the experiment would be done by the action group.

These sanctions were presented to the workers at two separate meetings. The first was conducted by the head of the development department and included an introduction to the socio-technical approach, reasons for the experiment, an announcement that the two departments might be combined into one, and the rules for sanctioning. Since this information was relatively new to the workers, they were given several days to assimilate it, and then a second meeting was held to answer questions, to make a decision, and to decide upon the composition of the action group. The consultants were present in both meetings to answer questions and to become acquainted with the workers. Surprisingly, few reservations about the experiment were voiced. Instead, workers seemed excited about trying something new and began, with little hesitation, to discuss the composition of the action group.

ACTION GROUP

The estimators and die engineers were given the opportunity to elect three workers from each department to represent them as members of the action group. This was done during the second sanctioning meeting described above. The supervisors from each department were also included, increasing the total membership to eight, in addition to two external consultants and a secretary. Because of heavy work demands, the action group was able to meet for only three hours per week, thereby increasing the length of the analysis and hypotheses generation phases of the experiment by several months.

SOCIO-TECHNICAL SYSTEM ANALYSIS

The socio-technical analysis was guided by an analytical model developed for use in service or advisory departments (see Appendix). Certain aspects of the model were modified to fit the existing situation, and an additional environmental analysis step was added. The stages of the model include: (1) general scanning, (2) determining the objectives of the system, (3) analyzing the roles in the system, (4) grouping the roles, (5) measuring the roles against psychological requirements, and (6) analyzing the environment.

General Scanning

This step provides a general introduction or problem orientation to the experimental system. The inputs, transformation processes, and outputs of the department are examined; its work roles, organizational structure, and geographical layout are also studied. This scanning provides a general background against which a more detailed investigation takes place.

The work load of the combined estimating and die engineering department may be broken down into four primary functions: (1) inquiries, (2) new orders, (3) revisions, and (4) currents.

Inquiries come into EST from sales and take the form of a request for an estimate to be used as part of a bid package for obtaining new business. Once a request for an estimate is received, a due date for the estimate is established with sales, and the estimate is officially logged-in and placed in a folder. Some estimates require a drawing of the finished product for the customer, and this is provided, in the form of a markup, by DE. After the markup is completed, the size and type of stock is estimated, and primary and secondary production processes are determined. Pieces per hour and percentage of expected scrap are also calculated. Weight calculations and material and shipping costs are then determined in addition to quality assurance requirements. The cost of dies is estimated and a final monetary value is placed on all previous calculations. A final estimate, including the markup, is then returned to the sales department.

New orders come into EST from sales. They are assigned an order number, and a delivery date is determined. The new order then goes to DE so that dies can be designed. Once in DE, the steel size is checked before the die shop orders steel for sinking a die. The new order is then placed on a die-design schedule and assigned to a die designer. The die designer makes a part drawing or uses the original markup and then designs the die in the form of a die layout which is used in the die shop as a blueprint for sinking a die. All die designs are checked by another die designer and then given to the die shop.

Revisions involve any customer changes in current forging products. They usually take the form of a request from the customer for cost adjustments related to changes in the original order. Requests enter EST from sales and follow the same basic form as a new inquiry. Again, DE may be involved if a revision markup is required.

Currents are a request for an undated estimate on existing business. They are used to determine a sales price for future orders (beyond the original order) on current or inactive business. Again, requests enter from sales and follow the same basic pattern as new inquiries. Table 1 outlines the basic inputs, transformations, and outputs of the EST/DE department.

The organizational structure consists of two subunits that correspond to the estimating and die engineering functions (remember that both subunits were originally different departments). Each subunit is headed by a supervisor who reports directly to the manager of the development department; he, in turn, reports to the works manager. The estimating function is structured according to the type of metal worked on—aluminum or titanium. The aluminum unit consists of two senior estimators and one estimator, while the titanium group has one senior estimator and two estimators. The remaining work roles deal with both types of metals and include two hand-forging and rings estimators, one weight estimator, a specifications clerk, a routing clerk, a senior secretary, and a part-time clerk typist. The die engineering function is structured according to the level of skill and complexity required in designing dies and includes (in decreasing level of skill): nine designers, two draftsmen, a detailer, a duplicating machine operator, a part-time clerk typist. Figure 1 displays the organizational structure of the EST/DE department.

The geographical layout of the EST/DE department is shown in Figure 2. The office space is located on the top floor of an older brick building which is used for the general offices of the X Company. Florescent lights and heating and ventilation conduits are suspended from an open-beamed ceiling which is painted to simulate a more natural ceiling. The noise level is quite low because of the quietness of the estimating and die engineering technologies (e.g., desk calculators, pencils, erasers, etc.). Visual obstruction is minimal except for the presence of pillars and files. The estimating and die

Table 1

INPUTS, TRANSFORMATIONS, AND OUTPUTS OF EST/DE DEPARTMENT

Input	From	Transformations	Output	To
Inquiry	Sales	Log in and place in folder; if markup needed, send to DE; estimate size and type of stock; determine production process; calculate pieces per hour, percentage scrap, weight, material and shipping costs; determine quality assurance requirements; and determine final money value of previous calculations.	Estimate with markup	Sales
New Order	Sales	Assign order number and determine delivery date; check steel size and inform die shop; place on die design schedule and assign to die designer; make part drawing and die layout; check.	Part drawing and die layout prints	Die shop
Revision	Sales	Same as inquiry	Revised estimate and markup and/or revised die layout and part drawing	Sales and die shop
Current	Sales	Same as inquiry	current estimate	Sales

Figure 1

ORGANIZATIONAL STRUCTURE OF EST/DE DEPARTMENT

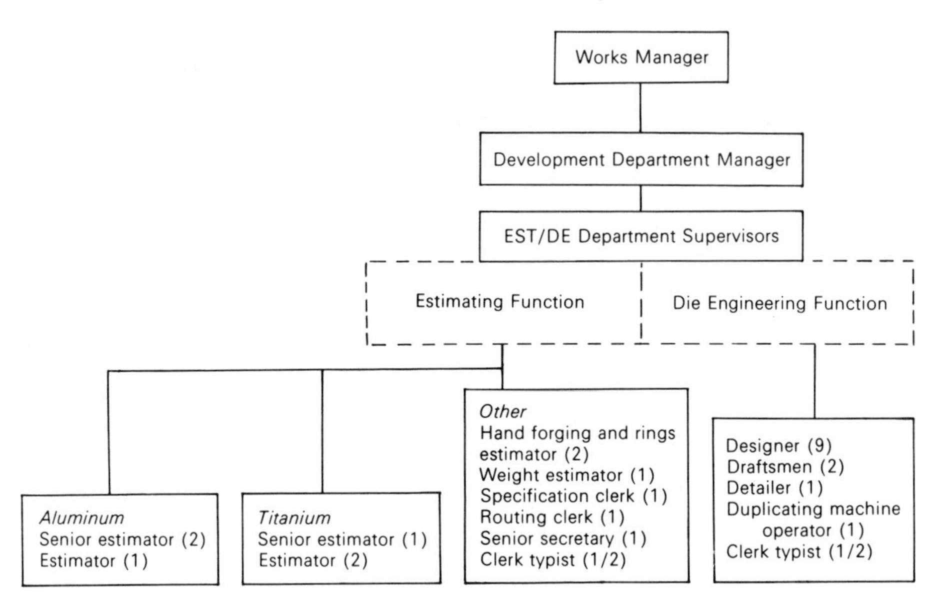

engineering functions are located adjacent to one another in a relatively open area partially divided by the placement of files, pillars, and tables. The location of desks in the EST unit corresponds somewhat to the natural task groupings of the estimating function, while the drawing tables in the DE unit are located at right angles to the overhead lights to eliminate the presence of bothersome shadows on the drawing boards. The supervisors' offices are located adjacent to the work area and are partially enclosed in glass which permits observation of most employees.

The Objectives of the System

Against the background of the initial scanning, it is now possible to do a more detailed analysis. In order to understand the activities of the EST/DE department, it is necessary to arrive at a clear definition of objectives, since this provides a standard against which to judge the appropriateness of workers' behavior. This is often a difficult task because objectives may not be formulated clearly or may be so abstract as to be of little use in directing behavior. A first step is to list all major outputs of the department. To be sure that no significant outputs are missed, it may be necessary to identify all inputs and trace them through their various transformations into outputs. Second, all outputs are examined for their appropriateness as objectives by sending them to the manager of the next higher-level system to see if these meet his demands. Since other departments often use these outputs, it may be necessary to enlist their help in determining which outputs are necessary objectives. Finally, since many outputs have meaning only in relation to an overall task or decision made outside the boundaries of the work system, it is often necessary to indicate their contribution to such decisions.

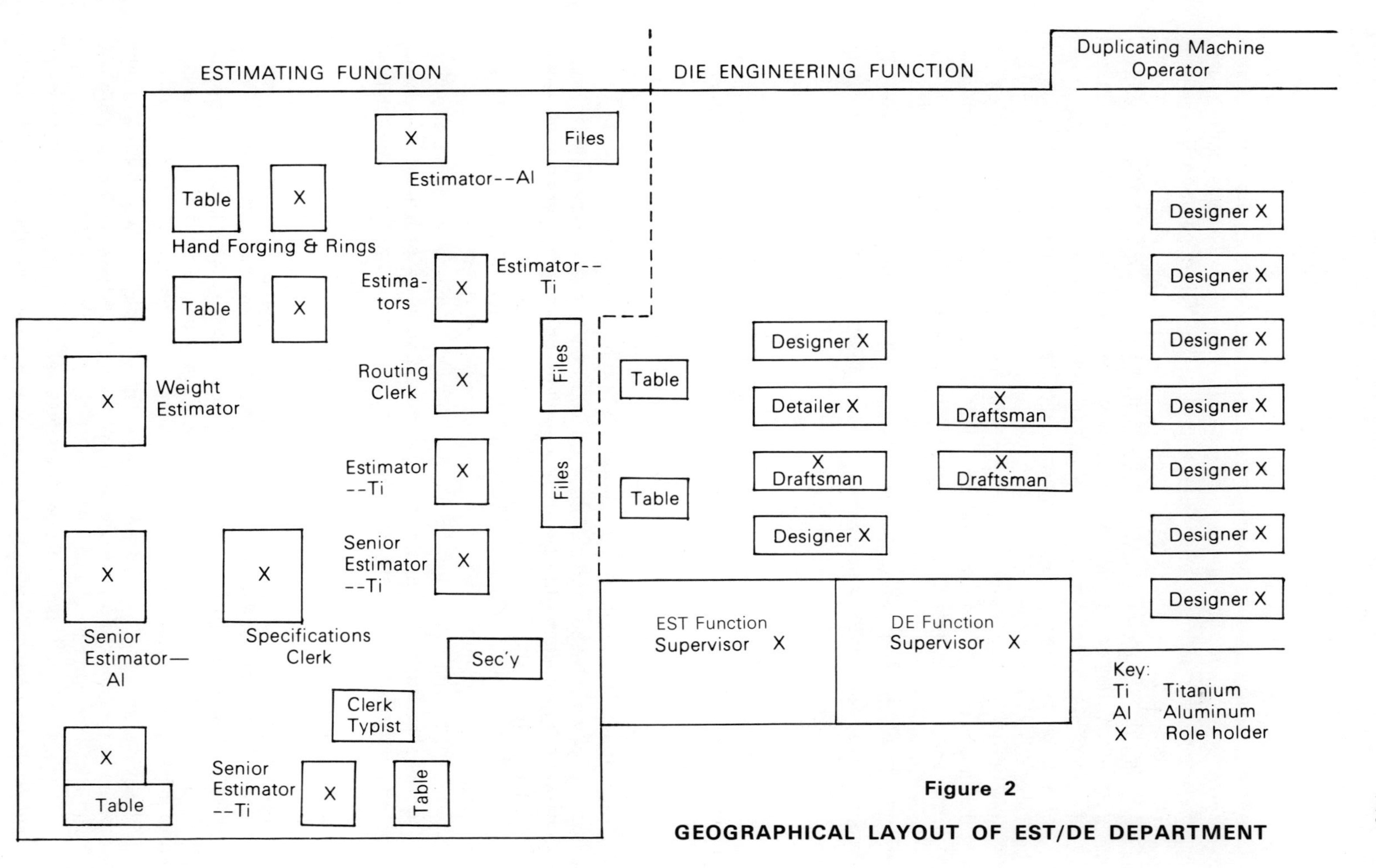

Figure 2

GEOGRAPHICAL LAYOUT OF EST/DE DEPARTMENT

To determine the objectives of the EST/DE department, a list of all inputs was generated. Then each input was traced through a transformation or conversion process until it became an output of the department. These outputs were listed as potential objectives and sent to the head of the development department for confirmation. A similar list was sent to the managers of each department that used the EST/DE department's services. This served as an additional check against possible errors or omissions. A final list of objectives was then described in terms of their intended destination, the overall decision to which they were intended to contribute, their required contribution to the overall decision, and the consequences of substandard performance with respect to social and economic costs. Table 2 examines the outcome of this analysis. The outputs are listed by functional unit—estimating and die engineering—since, at the time of this analysis, each unit viewed its objective separately. Interfaces between the two units are denoted with an asterisk to highlight internal interdependencies.

The output/objective analysis reveals the difficulty in trying to determine objectives for white collar work units. First, the number of objectives is quite large—11 for the estimating function and 17 for the die engineering function. Second, there exists no clear criteria for judging the relative importance of each output. Workers responded to this issue by stating: "We do everything that is required of us with as much accuracy as is possible." Third, there are no relevant measures of objectives. Because of the variable nature of each piece of work, counting the number of outputs is misleading because, for example, one estimate is not equivalent to another in terms of complexity, information required, clarity, etc. (It is interesting to note that this problem permeates the entire forging industry which is essentially a job-shop technology with 200 to 300 different products always in process.) Finally, outputs vary in relation to inputs from other departments. Since these inputs are also variable, workers are continually forced to balance their work distribution over the entire range of objectives. This is a difficult task given the lack of clear priorities for assigning work.

The nature of departmental objectives led to several problems affecting workers' performance. First, because of the difficulty of measuring objectives, workers had few clear standards against which to judge their performance. An inability to relate job behavior to clear measures of output resulted in few opportunities for on-the-job learning. Instead, workers were often left to their own judgment as to what constituted satisfactory work. Second, since the outputs were used by a variety of other departments, workers were dependent upon others for feedback about results. Since the results were often ambiguous and difficult to interpret, feedback was quite slow or non-existent. When it did occur, it was usually in the form of a generalized complaint which was difficult to link to actual job behavior. Workers felt that they were often blamed for things beyond their control, and that when specific feedback did come, it was weeks or months after-the-fact and of little use for correcting mistakes.

Finally, since many of the objectives were part of an overall decision made in another department, it was difficult to trace the affect of the EST/DE

Table 2

EST/DE DEPARTMENT OUTPUT/OBJECTIVE ANALYSIS

*Interface between EST and DE

Estimating Function

	Output	To Whom Sent	Overall Decision It Contributes Towards	Required Contribution to Overall Decision	Consequences of Sub-Standard Performance
(1)	New inquiry estimate	Development, sales	Bid package for new business	Helps determine sales price for new business by listing expected cost of producing product	Social: Lowers other people's confidence of estimating people Economic: Loses new business or loses money or new orders
(2)	Revision estimate	Sales	Sales price for changes to current business	Helps determine cost changes when customer requests changes in product	Social: Same as (1) Economic: Loses money on customer changes
(3)	Current estimate	Sales, routing, industrial engineering, accounting	Sales price for future orders on current business	Helps determine cost of producing current product in the future	Social: Same as (1) Economic: Loses money on future business
(4)	Change of procedures	Routing, planning, industrial engineering, process engineering *Die engineering	Evaluation of most economical fabrication procedure	Helps determine cost of changes in fabricating procedure	Social: Could stifle creativity Economic: Loses money or business by not being competitive
(5)	New order paperwork	Same as (4)	—	Initiates paper work to start new order into process to meet customer's commitment	Social: Same as (1) Economic: Loss of opportunity for receiving future business

Table 2 (*Contd.*)

Estimating Function (Contd.)

	Output	To Whom Sent	Overall Decision It Contributes Towards	Required Contribution to Overall Decision	Consequences of Sub-Standard Performance
(6)	Deck 8's	Data processing	Determination of standard costs	Helps keep standard costs up-to-date	Social: Same as (1) Economic: Loses money on present or future business
(7)	Metal transfer costs	Planning, accounting, industrial engineering, quality assurance	Establish firm transfer cost	—	Social: Same as (1) Economic: Same as (6)
(8)	Computer programs	Sales, industrial engineering, accounting, data processing	Faster and more accurate information	Develops and maintains existing computer programs as they pertain to estimating	Social: Same as (1) Economic: Same as (6)
(9)	Field inspection reports	Quality assurance	Accept or reject a recommendation	—	Social: Little effect Economic: Same as (5)
(10)	Cost evaluation, supplemental tooling, job costs	*Die engineering, process engineering, industrial engineering, quality assurance, production, sales management, planning	Identify major cost areas and evaluate changes	Helps determine costs for job and potential job changes	Social: Stifles creativity Economic: Adverse effect on costs, same as (4)
(11)	Ideas and information	*Die engineering, industrial engineering, process engineering, production	—	—	Social: Same as (1) Economic: Unknown

Table 2 *(Contd.)*

Die Engineering Function

Output	To Whom Sent	Overall Decision It Contributes Towards	Required Contribution to Overall Decision	Consequences of Sub-Standard Performance
(1) Markup	*Estimating, sales	Bid package for new business	Graphic representation of forging as shipped; complete enough to indicate required manufacturing operation	Social: Lowers other people's confidence of die engineers Economic: Could mislead customers, increase estimating time and decrease accuracy of costs
(2) Part drawing	*Estimating, sales, customer, die shop, quality assurance, production, process engineering, machine vendor	Substantiate quote, die sinking, verify specifications, inspection, blocker design	Verification of estimate by providing an accurate, scaled, and complete drawing	Social: Same as (1) Economic: Accuracy of die sinking, dimensional integrity of forgings, accuracy of blocker design
(3) Blocker drawing	Die shop, production, process engineering	Verify stock size, sink dies, design preforms	Providing information for sinking dies; accurate drawing depicts blocker design that will satisfy finish operation requirements	Social: Same as (1) Economic: Number of operations, accuracy of die duplication; too much scrap
(4) Hot finish drawing (cold work job)	Die shop, production, process engineering	Die sinking, blocker design, inspection of forgings	Accurate and complete drawing depicting required hot finish design which will satisfy cold work percentages and clearance requirement	Social: Same as (1) Economic: Same as (3)

Table 2 (*Contd.*)

Die Engineering Function (Contd.)

	Output	To Whom Sent	Overall Decision It Contributes Towards	Required Contribution to Overall Decision	Consequences of Sub-Standard Performance	
(5)	Die sinker's drawing and die layout drawing	Die shop, process engineering	Die sinking and die set-up	Provides special design information too extensive to be included in the part	Social: Economic:	Same as (1) Accuracy of die sinking; under stress on equipment; set-up time; dimensional integrity of forging
(6)	Die details, drawing and assembly drawings	Die shop, process engineering	Die sinking and die set-up	Provides detailed drawings of die components too complicated to describe on part drawing and die layout drawing	Social: Economic:	Same as (1) Same as (5)
(7)	Preform drawing	Production, process engineering	—	Provides a physical shape for stock which will reduce forging defects in subsequent die operations	Social: Economic:	Same as (1) Number of forging operations; cost of repair; cost of manufacturing preforms
(8)	Hand forging drawing	*Estimating	Operations, stock size and hand forging cost estimate	Provides detailed drawing showing hand forging dimensions, tolerances, specifications, starting stock size and operations	Social: Economic:	Same as (1) Cost of manufacturing and dimensional integrity
(9)	Drawing change notice form	Die procurement, routing, quality assurance	Record keeping, determination of die sinking requirements, forging line-up and inspection devices	Accurate record of drawings, changed and concise description of effect of change on specific dies and forgings	Social: Economic:	Same as (1) Delivery of dies and forgings; downtime of equipment; scrap if dies have not been changed

Die Engineering Function (Contd.)

	Output	To Whom Sent	Overall Decision It Contributes Towards	Required Contribution to Overall Decision	Consequences of Sub-Standard Performance
(10)	Die condition report and marked prints	Die procurement, process engineering	Die sinking	A list of instructions for revising existing die equipment	Social: Same as (1) Economic: Dimensional integrity of forgings
(11)	Customer revision form	*Estimating	Estimate revision, die cost and forging selling price	A concise description of the effect of a customer's revised print on dies and forgings	Social: Same as (1) Economic: Accuracy of customer changes on dies and forgings
(12)	Request for approval letter	Sales	Acceptability to customer of forging made to print	Explanation of differences between X Company's drawing and customer's drawing	Social: Same as (1) Economic: Possible rejection of forging by customer
(13)	Green form letter	Sales	Acceptability of customer's print	A list of differences between X Company's print and customer's print	Social: Same as (1) Economic: Future checking time same as (12)
(14)	Red and white forms	Sales	Effect of customer's print	Statement indicating no effect	Social: — Economic: —
(15)	Tryout report forms	Process engineering	Die tryout	A list of die operations scheduled for a new order	Social: — Economic: —
(16)	Machining cost information	*Estimating	Tooling and forging cost	Estimated costs and lists of operations and tools	Social: Same as (1) Economic: Inaccurate estimated cost-loss of revenue
(17)	Checked fixture and template drawing	Quality assurance	Acceptability of forgings	Drawings of required device plus pertinent instructions	Social: Same as (1) Economic: Same as (10)

department's contribution to the outcome of the larger decision. A new inquiry estimate and markup, for example, were part of a larger bid package generated by sales to acquire new business. Since sales could bid above or below the estimated price depending on the overall sales strategy, the affects of the actual estimate on winning or losing new business were virtually unknown. Estimators could blame the salesmen for asking too much or too little in regard to their estimate, and the salesmen could blame the estimators for not creating a competitive estimate. The circularity of this argument is obvious.

In summary, the analysis of the objectives of the EST/DE department revealed three major problems affecting the rational functioning of the work system. First, the system's objectives, though clearly related to outputs, were not formulated to provide workers with criteria for judging their relative importance. This made it difficult to determine task priorities and to estimate how much effort to expend on a particular output. Given a heavy work load, the lack of clear task priorities resulted in an unbalanced utilization of resources. Second, each objective represented a relatively unique piece of work making it difficult to determine standard measures of performance. Without this knowledge workers were unable to assess the results of their behavior. Third, the objectives contributed to decisions made in other departments. This intensified the feedback problem, since information about the consequences of these outputs was ambiguous and slow.

If these data are interpreted in terms of regulation and control, two necessary conditions for goal-directed behavior were absent: (1) a clear set of steady state variables (objectives) to serve as standards of performance and (2) timely and relevant feedback to inform workers of the state or condition of these variables. In the absence of these parameters, workers could not learn from their past behavior; furthermore, the ambiguous and variable nature of the feedback process left workers with the feeling of being often blamed for outcomes beyond their control.

Analyzing the Roles in the System

This step in the model is similar to the previous stage: the inputs, conversions, and outputs of each role are examined to arrive at role objectives. Once role objectives are determined, they are checked against the departmental objectives so that possible incongruities can be identified. It is important to note that the departmental objectives provide parameters for, and give meaning to, role objectives.

Each role in the EST/DE department was examined in regard to inputs, transformations, and outputs. Although the data are too numerous to list, here are some of the more important findings. The role analysis for the estimating function followed the outline of Table 2. Each role was listed in regard to outputs or objectives, to whom sent, overall decision contributed to, required contribution to overall decision, consequences of sub-standard performance, and the departmental objective it related to. Each EST role (11 in total) was examined with the exception of the half-time clerk typist; 32 role objectives were identified, and each was traced to one or more of the departmental objectives previously identified.

The role analysis for the DE function did not fit easily into the format used for the EST function. Each role holder did a variety of drawings depending upon the balance in the overall work load. Work assignments varied depending upon the complexity of the drawing and the abilities and skills of the role holder. Therefore, a simple input-transform-output analysis was not appropriate given the variability in work assignments over relatively short time periods, from a few hours to a week. Instead, each type of drawing was identified and related to an overall departmental objective (some drawings were departmental objectives); then, each role holder was listed in regard to the type of drawing he would be assigned depending upon his skills and the complexity of the drawing. This analysis produced a rank ordering of role holders based on skill level for each type of drawing. It was then possible to determine the type of drawing a particular role holder was likely to work on as well as the likely order of work assignment among the different role holders. This analysis identified 22 types of drawings (objectives) that were traced to one or more of the departmental objectives; each role (12 in total)—with the exception of the duplicating-machine operator and half-time clerk typist—was rank-ordered under one or more drawings, producing rankings that ranged from 2 to 9 role holders for each type of drawing.

The analysis of roles in the department produced the following information:

1. Roles in the EST function were structured according to the type of estimate, part process (weights, specifications, etc.), and/or metal (aluminum or titanium) worked on. Those in the DE function were arranged according to the complexity of the drawing and the skill level of the role holder.
2. Inputs to each role holder in the EST function varied over the entire range from simple to complex, while those for each role holder in the DE function tended to cluster around his skill level.
3. On the whole, task variety tended to be higher in the EST function than in the DE function.
4. Only a few outputs or objectives of the role holders in both functions were clearly measurable as standards for performance.
5. Feedback of results to a role holder was quicker and more effective within the department than across it.
6. Since the outputs of the roles contributed to decisions made outside the boundaries of the department, few role holders were able to observe the results of their efforts.
7. There appeared to be no incongruities between departmental objectives and individual role objectives.

Grouping the Roles

In order to identify relevant role interactions, it is necessary to examine the grouping of roles in relation to the current work process. Once the existing interdependencies are determined, we can begin to hypothesize role relationships or clusters required for task performance. The outcome of this analysis provides information as to how well the current departmental structure matches task requirements; it also indicates potential social problems that might arise if the role relationships are changed.

To determine role interaction links in the EST/DE department, the steps of each departmental transformation process were listed in sequential order, and the different roles responsible for each step were identified. In the EST function, the transformation stages for new inquiry, revision, current, and hand forging and rings estimates were listed for both aluminum and titanium metals. This produced role groupings that ranged from three roles (hand forging and rings estimate) to seven roles (new inquiry estimate, titanium). Table 3 lists a typical grouping-of-roles analysis for a new inquiry estimate in aluminum. One can see that the transformation process entails nine stages that are carried out by six different role holders in addition to one boundary crossing into the DE function for a markup. It is interesting to note that this transformation process resembles a long-linked technology, since it is sequentially divided into smaller units that are carried out by different role holders. A division of labor work design is evident.

In the DE function, the transformation stages for each type of drawing were listed, and the role holders responsible for each stage were identified. This analysis revealed a rather simplified grouping of roles that included the following stages: (1) drawing assigned by supervisor to a designer, draftsman, or detailer; (2) drawing completed by designer, draftsman, or detailer; (3) drawing checked by another designer or draftsman; and (4) drawing returned to supervisor if no mistakes are found, or if mistakes are found steps (2) and (3) are repeated. The grouping of roles between the person doing the drawing and the individual checking it is determined by the complexity of the drawing and the skill level of the role holder; the checker is always of an equal or of a higher skill level than the drawer.

The grouping of roles analysis revealed the existing role interactions in the EST/DE department. Depending upon the type of work, the transformation stages involved three to seven different roles in the estimating function, and two to three different roles in the die engineering function. Each trans-

Table 3

GROUPINGS OF ROLES ANALYSIS: NEW INQUIRY ESTIMATE, ALUMINUM

Transformation Stages	Work Role
1. Stand-up meeting	EST function supervisor; estimator
2. Log-in and forward	Senior secretary
3. Mark up	Done in DED function
4. Line-up	Senior estimator, I
5. Die estimate	Senior estimator, I
6. Weight and specifications	Weight estimator
7. Computer extension	Clerk typist
8. Extension check	Senior estimator, I
9. Log-out, forward, obtain delivery	Senior secretary

formation process entailed a sequential ordering of tasks through which paperwork flowed from one role to another. Role interactions between the EST and DE functions were also examined to ascertain the extent to which work in one functional area was dependent upon work in the other area. New inquiry estimates required markups from DE, while new orders required initiation in EST before they could be processed in DE; numerous types of job-related information also crossed the internal boundary. Task requirements for EST included gathering a diversity of information and applying it correctly to each estimate. The division of labor design principle is questionable, since overall task performance requires a good deal of interaction among role holders. This method of work design also increases the probability that relevant information will be lost as the paperwork passes from one role to another. Task requirements for DE entailed detailed drawing in relation to customer specifications. The organizational design of a separate drawer and checker appears to meet these requirements. Because of the role interactions between the two functions, the separation of the department into EST and DE functions is also questionable. The geographical layout of the department (see Figure 2) appears to enhance this functional split. A potential problem that could limit the role interaction between functional units is the historical split between estimators and die engineers as well as skill and status differences between each function.

Measuring Roles against Psychological Requirements

To ascertain the extent to which each role meets basic psychological requirements, it is necessary to learn the workers' perceptions of their own work roles. This may be done by an interview or a questionnaire. Emery (1963) has developed a general set of psychological requirements that pertain to the content of a job; these may be used to determine the extent to which the psychological requirements are presently met on-the-job. The list of requirements includes:

1. The need of the content of a job to be reasonably demanding in terms other than sheer endurance and yet providing a minimum of variety (not necessarily novelty).
2. The need to be able to learn on the job and go on learning.
3. The need for some minimal area of decision-making that the individual can call his own.
4. The need for some minimal degree of social support and recognition in the workplace.
5. The need to be able to relate what he does and what he produces to his social life.
6. The need to feel that the job leads to some sort of desirable future (not necessarily promotion).

The measurement of the roles against the psychological requirements for the EST/DE department was done in two phases. The first stage involved the construction of a semantic differential (SD) instrument in which nine bi-polar scales were used to measure the psychological requirements; nine additional scales were listed to measure other aspects of workers' jobs including evaluative attitudes. Figure 3 displays the SD instrument. Each role holder was asked to rate the concepts "my job" and "my ideal job" on each bi-polar scale with six intervals separating the poles. The responses to the target "my job" were then subtracted from the responses to the concept "my ideal job," and the absolute difference between the two concepts was used as a measure of psychological need deficiency. This analysis identified those roles in which the psychological requirements were deficient; it also revealed the actual dimensions (scales) which caused the mismatch.

Figure 3

SEMANTIC DIFFERENTIAL INSTRUMENT: PSYCHOLOGICAL REQUIREMENTS

Concept being rated: *"My job as it currently exists" and "My job as it should ideally exist."*

LEFT-HAND POLES	Highly	Moderately	Slightly	Slightly	Moderately	Highly	RIGHT-HAND POLES
Provides for a variety of tasks*							Does not provide for a variety of tasks
Rigidly structured**							Loosely structured
Does not require skill**							Requires skill
Pleasant***							Unpleasant
Provides for learning*							Does not provide for learning
Non-essential**							Essential
Does not provide for decision-making*							Provides for decision-making
Valuable***							Worthless
Requires speed**							Does not require speed
Does not provide for status and recognition in the department*							Provides for status and recognition in the department
Provides for setting standards of quality and/or quantity*							Does not provide for setting standards of quality and/or quantity
Boring***							Interesting
Does not provide for feedback of knowledge of results*							Provides for feedback of knowledge of results
Requires accuracy**							Does not require accuracy
Will not have more importance in the future**							Will have more importance in the future
Leads to a desirable future*							Does not lead to a desirable future
Not well regarded by others*							Well regarded by others
Provides a contribution to the overall decision to which it was intended*							Does not provide a contribution to the overall decision to which it was intended

* Measures a psychological requirement
** Measures appropriate to EST and DE functions
*** Evaluative or attitudinal measures

The second phase of the measurement procedure involved an in-depth interview in which responses to the SD instrument were examined in more detail. The interview was conducted with those role holders whose SD data revealed large deficiencies (40 percent or more) or whose responses were difficult to interpret; the interview was also extended to anyone in the department who wished to provide more information. It was anticipated that the interview data would provide a separate measure of the degree to which each role met the psychological requirements.

The data from the nine bi-polar measures of psychological requirements were examined to determine the extent to which workers' perceptions of their "ideal job" differed from those of their "present job." Each role was inspected scale by scale, and each time the "ideal job" differed from the "present job" by 40 percent or more, a deficiency was recorded. A summary of the results of this analysis appear in Table 4. The number of roles in which deficiencies occurred and the percentage of total are listed for each psychological requirement for both departmental functions. The data reveal a striking similarity in the percentage of roles which have deficiencies on the same psychological requirements for both functions. This is especially evident when one examines those scales on which deficiencies occur for over 30 percent of the role holders: "provides for status and recognition in the department," "provides for feedback of knowledge of results," "leads to a desirable future," and "well regarded by others." Although there is no standard against which to compare these results, there appears to be ample room for improving the degree to which the present roles meet the psychological requirements.

Table 4

SEMANTIC DIFFERENTIAL MEASURES OF PSYCHOLOGICAL REQUIREMENTS

Summary of Number of Roles in which "Ideal Job" Differed from "Present Job" by 40 percent or more*

	EST Function (N = 12)		DE Function (N = 16)	
Psychological Requirements	No.	% of Total	No.	% of Total
1. Provides for variety of tasks	2	16.6	1	6.3
2. Provides for learning	3	25.0	3	18.7
3. Provides for decision-making	1	8.3	2	16.6
4. Provides for status and recognition in the department	4	33.3	7	43.8
5. Provides for setting standards of quality and/or quality	1	8.3	2	12.5
6. Provides for feedback of knowledge of results	6	50.0	9	56.3
7. Leads to a desirable future	6	50.0	7	43.8
8. Is well regarded by others	5	41.7	7	43.8
9. Provides a contribution to the overall decision to which it was intended.	2	16.6	1	6.3

*The "ideal job" responses exceeded the "present job" responses for all data.

The interview data provided further insights about the SD responses. Interviewees reported that the primary reason why they felt that their roles did not provide for status and recognition in the department and were not well regarded by others was the lack of feedback of results; and when feedback was received, it was usually negative. They reported that "We always get blamed when things go wrong"; "They only let us know when we make mistakes, never when we do something right." The feeling that their jobs did not lead to a desirable future was explained in terms of the "dead-endedness" of the EST/DE department. Little turnover, few chances for transfers, and a low probability for promotion to management combined to give the workers a sense of resignation to their position in the company. In general, the interview data supported the SD data in regard to the degree that the psychological requirements were presently met.

The psychological need data may be interpreted in light of the preceding analysis. Specifically, the departmental objectives and individual role data revealed a problem in relating task behavior to objectives and in receiving feedback of results. One consequence of this dysfunctional regulatory process was that workers were frequently blamed for problems beyond their control. Feelings of low status and recognition and of low regard by others appear to result from this inadequate control structure. Without clear standards of task performance and timely and relevant feedback, workers are exposed to the subjective and often arbitrary assessment of others. Here, we have rediscovered a major point in socio-technical analysis: the social and psychological aspects of work cannot be understood apart from the detailed facts of the task system and how it operates in an actual work situation. In this instance, workers' needs for status and recognition are severely constrained by the task system's control structure.

Analyzing the Environment

Since socio-technical systems continually exchange materials and information with their environment, it is necessary to examine relevant environmental interfaces to ascertain the extent to which these exchanges meet system needs and environmental demands. This analysis is carried out by listing all inputs that enter the system and all outputs that leave it. The inputs and outputs are then traced to their source or destination, and each identified interface is examined for discrepancies between inputs and department needs or between outputs and environmental demands. With these data, it is possible to determine the degree to which the system is matched to its environment.

The environmental analysis for the EST/DE department proceeded in two steps. First, all inputs and outputs were examined in regard to their source or destination. This allowed for an identification of all relevant exchanges with other departments in the company. Figure 4 portrays the results of this analysis. Second, to determine the extent to which these interchanges matched departmental needs or environmental demands, data were collected from 42 relevant role holders from the external departments (identified in Figure 4). Two types of data were gathered—in-depth interviews and responses to a semantic differential instrument. The interview questions included:

Figure 4

EST/DE DEPARTMENT: ENVIRONMENTAL EXCHANGES

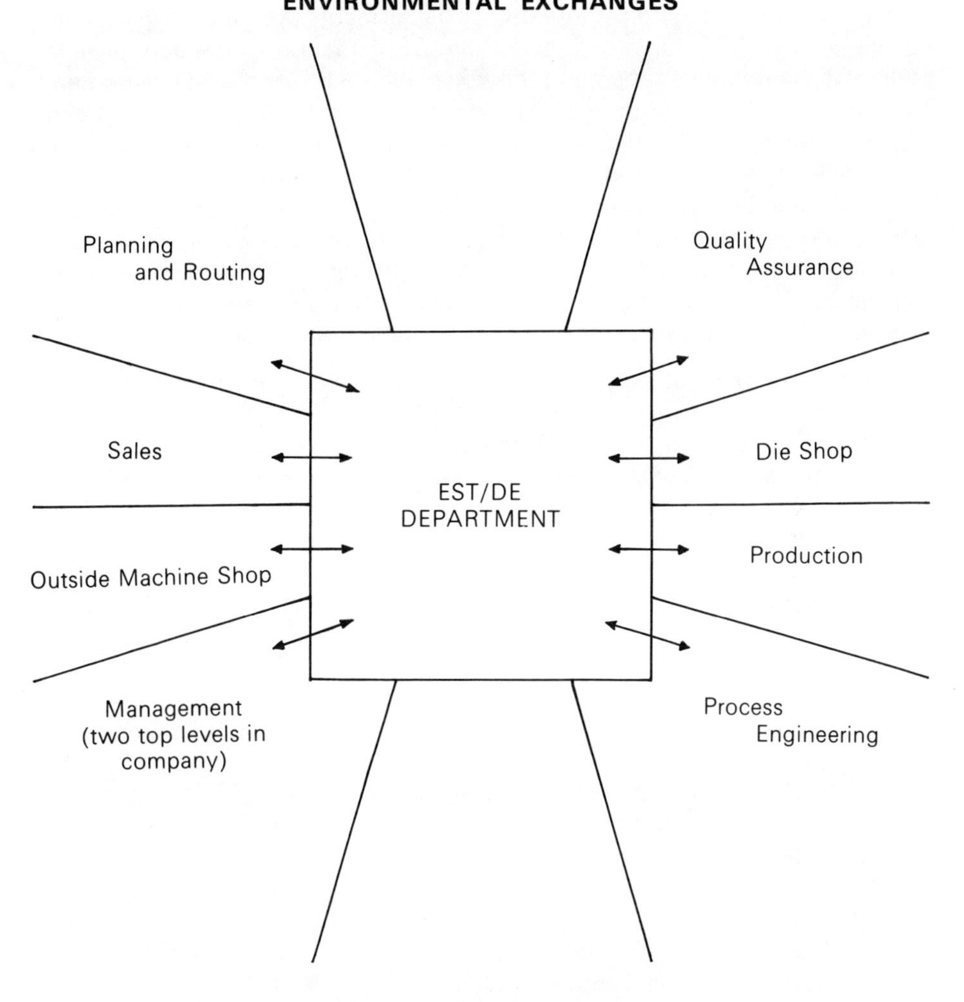

1. In what areas could the relations between your department and the EST/DE department be improved?
2. What information could you give the EST/DE department, that you do not already give, to improve their performance?
3. What information could the EST/DE department give your department, that they do not already give, to improve your performance?
4. What could the EST/DE department do to increase their effectiveness in regard to your department?
5. What could your department do to increase the performance of the EST/DE department?
6. How could the relations between your department and the EST/DE department be improved?

A SD instrument was constructed to obtain perceptions from

the external role holders as to how well the EST/DE department met environmental demands. Twenty-three bi-polar scales were used to measure different facets of the EST/DE department's activities and performance. Each member from the external departments was asked to rate the concepts "the EST function" and "the DE Function" on each bi-polar scale with six intervals separating the poles. Figure 5 lists the SD instrument. The SD data helped to determine the extent to which the EST/DE department met environmental demands and also identified the actual dimensions on which mismatches occurred.

The interview data provided a wealth of information for the environmental analysis. Responses to the six interview questions were quite specific as they were often given in technical terms. An extensive list of suggestions for improving interchanges with the EST and DE functions was generated from the data. Typical recommendations included:

1. "Communicate machine shop problems in regard to fixtures, templates, gauges, etc., to DE function."
2. "Get designers down on the shop floor periodically because of 'knowledge gap' between operating needs and die designs."
3. "Move EST function physically closer to sales."
4. "Provide a table of mechanical property minimums for standard products so EST function could go ahead and quote properties."

In addition to suggestions for improving interchanges, a variety of interface problems were identified. These included:

1. "Need quicker answer from DE function on supplemental tooling."
2. "Have trouble in gray area of finish die revisions when there is a die layout report involved, and especially when bending or cold work is a subsequent operation."
3. "Need better communication with EST function in specifications area."
4. "Estimators shouldn't be too forceful with their concepts—more moderation with thoughts."

The general tone of the interviews was optimistic. Individuals were willing to discuss problem areas and offer specific solutions. Although the tenor of the responses varied according to the departmental identity of the interviewee, the data strongly suggested that environmental interfaces could and should be improved.

The SD data provided a quantitative representation of external role holders' perceptions of the EST and DE units. Since there were too few data on which to perform a satisfactory factor analysis, the responses were interpreted in terms of each scale. Table 5 provides a summary of these data. To provide an overall view of the results, the responses were first averaged over all scales, and then a mean of these averages was computed for each external department. The results provide a crude index of how role holders from external departments perceive the functioning of the EST and DE units;

Figure 5

SEMANTIC DIFFERENTIAL INSTRUMENT: ENVIRONMENTAL ANALYSIS

Concept being rated: *"The EST or DE Function"*

LEFT-HAND POLES	Highly	Moderately	Slightly	Slightly	Moderately	Highly	RIGHT-HAND POLES
Valuable							Worthless
Slow							Fast
Not wasteful of resources							Wasteful of resources
Chaotic							Orderly
Essential							Non-essential
Obsolete							Current
Reliable							Not reliable
Not potentially more important to my department							Potentially more important to my department
Well regarded by others							Not well regarded by others
Destructive							Constructive
Capable							Not capable
Becoming weaker							Becoming stronger
Important							Not important
Not accurate							Accurate
Easy to get along with							Not easy to get along with
Not on time with work that is needed by my department							On time with work that is needed by my department
Non-threatening							Threatening
Needs improvement in relations with my department							Does not need improvement in relations with my department
Open							Closed
Not friendly							Friendly
Offers valuable advice to my department							Does not offer valuable advice to my department
Does not openly seek advice and suggestions							Openly seeks advice and suggestions
Complete with information							Not complete with information

Table 5

SUMMARY OF SEMANTIC DIFFERENTIAL RESPONSES

Means for External Departments Averaged
over 23 Bi-Polar Scales*

External Departments	EST Function	DE Function
Planning and Routing	81 (n = 46)**	88 (n = 46)
Sales	61 (n = 39)	88 (n = 69)
Quality Assurance	74 (n = 115)	73 (n = 115)
Production	68 (n = 92)	83 (n = 115)
Die Shop	61 (n = 46)	80 (n = 115)
Process Engineering	69 (n = 69)	79 (n = 184)
Outside Machine Shop	74 (n = 46)	86 (n = 92)
Management	63 (n = 161)	70 (n = 161)

*All responses were converted from six interval scores to scores ranging from 0–100 for easier interpretation $\left[\frac{(n = 1)}{.1}\right]$

**All means have been inflated by a factor of 23 due to collapsing all scales into one grouping.

for example, the sales and die shop departments rate the EST function moderately low (m = 61 and m = 61) and the DE function moderately high (m = 88 and 80). It is interesting to note that managers (top two levels in the company) rate both functions moderately low (EST m = 63; DE m = 70). The results also show that, except for one department (quality assurance), all external departments rate the EST unit lower than the DE unit. This is not surprising given the type of work done in each unit. Estimating by its very nature entails some degree of guessing; therefore, the probability of making mistakes and thus incurring the wrath of others is high. Die designing is a more precise type of work, and although mistakes are made, their frequency is quite low. The SD data support the interview conclusions in that there is ample room for improving environmental interfaces in the EST/DE department.

Summary of the Socio-Technical Analysis

Before moving to the next stage of the experiment, hypotheses for redesign, here are some of the conclusions drawn from the socio-technical analysis of the EST/DE department:

1. The EST/DE department exists primarily to serve other departments. It receives informational inputs from other departments and returns these in the form of estimates, drawings, and other documents that are used to carry out the work of the external departments. This support function places the EST/DE unit in a crucial position in the company: the EST function is needed to acquire new business, while the DE function is required to start the production of new business.
2. The geographical layout and workflow in the department

increases the division or separation between the EST and DE functions. Files and tables serve as physical barriers between the units, and different roles and functional identities provide psychological boundaries between the two subunits.

3. Departmental objectives are numerous and difficult to measure. Clear criteria for judging the importance of objectives do not exist. Workers do not have meaningful feedback of results, and they must rely primarily on their own judgments or the assessment of others as to how well they are performing. It is difficult to determine the consequences of sub-standard performance and to trace sources of mistakes. The ambiguity of their contributions to other departments leaves workers open to criticism.
4. Inputs are highly variable and novel. Without clear task priorities workers have a difficult time balancing their work loads and determining the time and effort required to deliver their outputs.
5. Role objectives are difficult to measure. Since these objectives or outputs contribute to larger decisions made outside the boundaries of the department, workers are not able to observe the overall results of their performance.
6. In the EST function, roles are structured according to the type of estimate, part process, or metal worked on; those in DE are structured according to the complexity of the drawing and the skill level of the workers. Task variety and opportunities for independent decision-making accrue to highly skilled job holders (designer and estimator).
7. Roles in the EST function are grouped according to the sequential order in which tasks are carried out for each departmental transformation process. This division of labor work design thwarts necessary task interactions and increases the probability that information will be lost in process. It also gives some roles a rather limited piece of a whole task.
8. Roles in the DE function are grouped according to the level of skill of the role holder. Since work is assigned according to skill level, highly skilled workers tend to do the most challenging work.
9. Workers perceive a difference between what "ideally should be" and what "presently exists" in regard to the degree that their jobs fulfill certain psychological requirements. This is especially evident concerning status, prestige, and recognition needs.
10. External departments depending on the EST/DE department for services feel that interface problems exist. Ambiguous feedback from external departments accounts for some of the internal feelings of low status and recognition.

HYPOTHESES FOR REDESIGN

Against the background of the socio-technical analysis, it was possible to hypothesize several ways to redesign individual jobs as well as the entire EST/DE department. As a first step toward this task, members of the action group split into small groups to generate alternative ways of organizing the estimating and die engineering tasks. To facilitate this process members were asked not to evaluate the various proposals until all alternatives were formulated clearly. Once an extensive list of hypotheses was formulated, each proposal for departmental redesign was examined using the following criteria:

1. Opportunities for psychological need fulfillment
2. Expected affect on objectives
3. Relationship to the problems identified in the analysis
4. Expected affect on the environment
5. Degree to which workers' trust and confidence would be maintained.

In contrast to the evaluation procedure used for departmental redesign proposals, the individual role hypotheses were listed for each worker and then saved until a final departmental design was determined. Since many of the departmental proposals involved some form of autonomous work group design, it was felt that workers could use the individual role hypotheses as data for organizing their own tasks within the autonomous work groups. This would give workers some choice in designing their jobs with the benefit of suggestions derived from the socio-technical analysis.

Although the hypotheses generation process started rather slowly, the final outcome was substantial. Seventeen departmental redesigns were suggested and evaluated. Examples include:

1. A sales engineering group composed of estimators and die engineers attached to sales, a product/process development group made up of estimators and die engineers attached to the development department, and a die design group comprised of die engineers attached to die manufacturing.
2. An engineering producibility group of estimators and die engineers that sets up information to be processed by a group of cost estimators and die engineers.
3. Autonomous teams of estimators and die engineers grouped according to type of business—military air, military other, commercial air, commercial other.
4. Autonomous teams of estimators and die engineers grouped according to types of forging equipment—mechanical presses, upsetters, hammers, and hydraulic presses.
5. A new business group of die engineers and estimators, an estimating group for maintenance and revision of existing business, and a die design group attached to die manufacturing.

The individual job hypotheses were also numerous totaling over 300 proposals for the 28 roles in the department. Examples of these suggestions include:

1. Draftsman: training program to up-grade worker to die design status.
2. Senior designer: provide cross-training in estimating procedures.
3. Designer: provide an opportunity to observe propeller production procedures in order to remain current.
4. Hand-forging estimator; provide cross-training in die engineering so he can design his own hand-forgings.
5. Weight estimator: train to check and evaluate customer revisions.
6. Clerk typist: set up training program and assign specific duties to role.

Although the criteria for evaluating departmental hypotheses were primarily qualitative, members of the action group were able to decide upon a final redesign. The department would be structured into five autonomous work groups according to the natural flow of work through the department. (1) A *new business team* (four estimators and four designers) would be responsible for generating estimates and markups for use by sales in obtaining new business. (2) A *start-up team* (one estimator and five designers) would coordinate all new orders generated. This would involve die designs, tryouts, and a liaison function between die manufacturing, process engineering, and production. (3) A *maintenance team* (three estimators and four die engineers) would maintain and improve existing business after the start-up team's duties were discharged. This would include currents, revisions, standard costing, and job monitoring. (4) A *support team* (the secretary, clerk typist, and duplicating machine operator) would provide secretarial, clerical, computer, and duplicating services to the other teams. (5) A *management team* (the two functional supervisors and a third supervisor who was responsible for certain die engineering duties) would provide technical and social assistance to the work teams and help them move toward responsible autonomy. The actual duties of this team were not defined clearly because of the uncertainty of how well the teams would function under the new design. Figure 6 displays the redesign in relation to the present hierarchy of the company.

Guidelines for autonomous group functioning were used to organize each work team. The guidelines and their anticipated use in the redesign included:

1. "Autonomous work groups should have clearly defined and easily measurable inputs and outputs"—It was anticipated that each work team would develop, in conjunction with management, a scheme for measuring and for setting priorities for its specific inputs and outputs. This would provide workers with appropriate task priorities as well as standards for evaluating performance. Teams would also devise, with the help of external departments, an appropriate feedback process.
2. "Autonomous work teams should have clearly defined physical, task, and social boundaries"—the work teams were

Figure 6

EST/DE DEPARTMENT REDESIGN

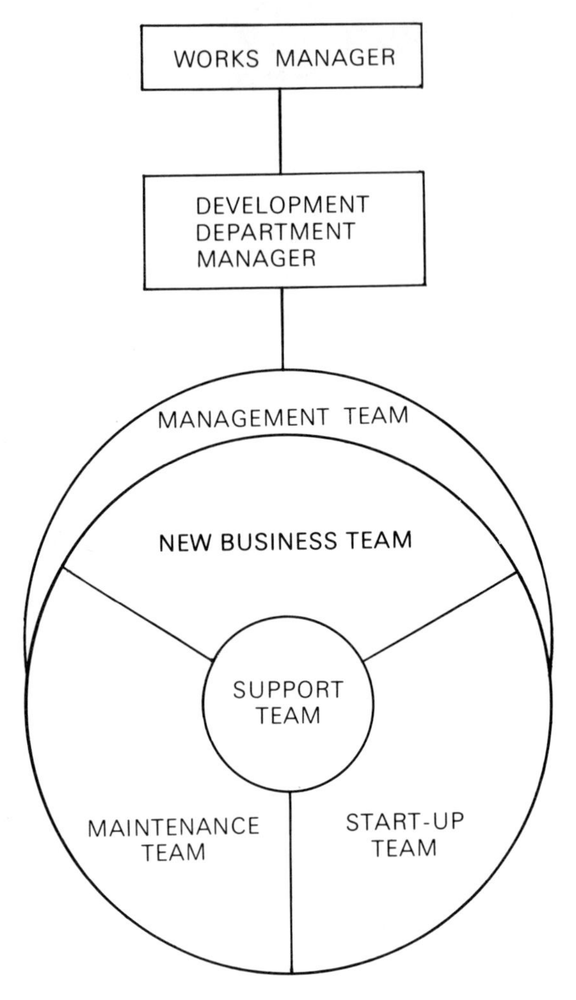

to move into clearly defined work areas that would enhance social interaction. The primary task of each work team was clearly defined and relatively independent from the other teams.

3. "Autonomous work teams should contain a requisite variety of skills for task accomplishment, group maintenance, and adjustment to change"—The role composition of each team provided somewhat the requisite task skills; however, additional training was needed for group maintenance skills and cross-functional skills to allow for internal regulation.
4. "Autonomous work group members should collectively take responsibility for goal attainment"—It was felt that the team structure would provide necessary task identity as well as clear boundaries of responsibility and discretion.

5. "The task of an autonomous work group should be relatively independent and form a self-completing whole"—Each work team was structured around a primary task that formed a relatively independent and self-completing whole.
6. "Autonomous work team members should see themselves as a team and should be perceived by others as a team"—Each team was to be involved in intensive team-building activities to promote a shared team identity. Environmental units were also going to be related to the new team structure. The physical redesign would also facilitate the formation of team identities.

The redesign was intended to overcome many of the problems inherent in the existing departmental structure. The basis of the autonomous group design was to organize the departmental tasks into three relatively independent groupings corresponding to the natural flow of work through the unit: acquiring new business, starting-up new business, and maintaining existing business. This design offered several advantages over the present structure which divided the unit into two task functions (EST and DE), and then arranged tasks further by skill level (DE function) or by past process and material worked on (EST function).

First, the team structure placed interdependent tasks and relevant role holders into relatively self-contained groups with clear physical boundaries. This would increase necessary task interaction and would reduce the distance over which information flowed. Second, the team design provided workers with a primary task identity in addition to well-defined task boundaries. This would give workers an opportunity to derive recognition from identifying with and performing a whole task, and it would provide clear boundaries for discretion and responsibility. Third, the group design allowed workers greater task variety and cross-skill learning as individuals from both functions (EST and DE) were required to manage a common task. Finally, the team structure gave workers the opportunity, with the help of management and other departments, to organize their own work roles, to decide upon task priorities, and to devise a relevant feedback process. This freedom would enhance workers' self-esteem and lead to an effective regulatory process. Given these advantages, the new work structure was expected to provide a needed match between the requirements of the EST/DE tasks and the needs of workers.

IMPLEMENTING THE REDESIGN

The transition from the existing departmental structure to the new redesign encountered more problems than were expected. Several members from the EST/DE department did not fully accept the new design; they had mixed reactions to the proposed team structure in addition to doubts about the need for a departmental restructure. Given these reservations, members agreed to try out the new design for a six-month period, after which a decision concerning its appropriateness would be made. Management agreed to experiment with the redesign; however, their support was based as much on the need to improve the performance of the department as on the actual merits of the redesign. Similar reactions were voiced by members from external departments. Here, responses to the autonomous team structure were varied, yet assurances were given to help the estimators and die engineers try something new.

Once agreement was reached to experiment with the new design, departmental members were anxious to move into the team structure. Because of vacations and a variety of other circumstances, a geographical relocation was postponed for six weeks. This somewhat dampened workers' enthusiasm as it was felt that a physical change would be a natural breaking-point between the old and the new design. After the physical restructure, individuals were faced with the problem of disengaging from tasks falling outside of their work team's boundary and engaging in new tasks; also, team members had to devise appropriate task interactions as well as ways of relating to other teams, to management, and to external departments.

Here, it is important to discuss two key organizational changes with a negative impact on the implementation process. First, the X Company was in a severe economic crisis; aerospace business which accounted for about 80 percent of the firm's sales was in a sharp decline, and the company was losing money for the first time in its history. This economic change placed heavy demands on the firm's ability to generate new commercial business, especially in those departments, such as EST and DE, that were responsible for new products. Given the current business situation, the protection necessary for socio-technical experimentation was relatively absent. Rather than provide workers with reduced work loads, management expected individuals to implement the redesign while generating and starting up new business. Under these circumstances, workers were faced with the almost impossible task of experimenting with organizational change without the necessary experimental protection. Second, a change in works managers disrupted the implementation process. During the analysis phase of the experiment, the works manager who initially sanctioned the experiment was transferred to another plant. The new works manager inherited the experiment with less than total support for this approach to organizational change. Rather than terminate the project, he decided to give it tacit approval under the assumption that it might lead to desired improvements. The company's decision to replace works managers left workers with ambiguous feelings about the firm's support for socio-technical experimentation. Although assurances were given, workers felt that management was not fully committed to this strategy. In retrospect, both organizational changes reduced considerably the sanction required for socio-technical experimentation. The full impact of these unexpected turns of events came sharply to focus during the implementation stage of the experiment.

To help the teams cope with emergent problems, weekly team meetings were held on a voluntary basis with external consultants. Issues such as work load balancing across teams, task assignments, team responsibilities, and leadership were disucssed. Workers were caught in the bind of having to face a variety of novel problems while still carrying out their normal work loads. First-level supervisors were confronted with a similar problem:

relating to teams of workers and helping them to move in the direction of responsible autonomy, while still fulfilling their regular managerial duties. To work on management/team interface issues, meetings were held for the purpose of deciding team and management responsibilities. Each team prepared a list of duties for which it would be responsible, and management prepared a similar list for each team. Then the lists were matched and incongruities were discussed. Although this process produced many disagreements between management and the workers, it provided team members with some opportunity for forming team identities.

Because of work load pressures, team members and management were not able to spend the time needed to work through their problems. This slowed the transition process and aggrevated earlier doubts as to the usefulness of the redesign. Time constraints also impeded progress toward meeting some of the more crucial requirements for autonomous group functioning: measurable inputs and outputs, feedback of results, requisite task skills, and clear group boundaries. In the absence of these requirements, many of the problems from the old structure were carried over into the new design. In addition, some members from external departments complained that outputs from the work teams were inferior and took longer to produce than outputs from the old structure. Although there were no clear data to support these accusations, they had a negative effect on the workers.

In spite of the transition problems, workers were able to keep up with much of their work load while attempting to solve some difficult issues. Although the team structure was difficult to launch, workers made valid attempts to make it function. To facilitate this process and to give the experiment a needed assist, a resanctioning process was carried out. The outcomes of this process included:

1. Agreements to hold weekly team meetings for the purpose of increasing team skills and solving task related problems.
2. Agreements by management to help each team meet the minimal requirements for autonomous group functioning.
3. Agreements from some environmental units to reduce pressures on the experimental units.
4. Agreements to work on a new reward structure to match the team structure.
5. A renewed patience with the problems and time involved in making a new design work.

The resanction provided workers and management with an important reality base as to the work involved in implementing the new structure. Although this resulted in some support for the experiment, two major problems continued to disrupt team functioning: insufficient task skills within teams and inadequate feedback of results. The first problem involved the skills necessary to perform each group's primary task. Although members of each team possessed many of the requisite skills, it was expected that additional skills would be learned during the implementation process. Because of heavy work demands, individuals did not have the time to develop new skills; this restricted the allocation of tasks among group members in order to balance their work loads. To overcome this difficulty, team members were forced to rely on individuals from other teams to perform some of their tasks. This resulted in uneven task allocations across teams as well as a disruption of primary task identity. In short, the membership boundaries of each group were subject to unexpected variations in personnel depending upon the skills needed for team tasks. Given this problem, it was difficult to develop permanent work teams with primary task identities.

The second problem concerned workers' knowledge about the results of their performance. Again, it was anticipated that team members would be able to devise appropriate measures of their outputs as well as a feedback process to inform them of deviations from these standards. Given the pressure to generate new commercial business, workers were subject to an almost continuous input of novel work. The uniqueness and variability of this work precluded clear knowledge of what was acceptable performance; that is, it was difficult to develop acceptable standards of performance for work that was unique to the company. Furthermore, external departments, such as sales, were confronted with similar problems associated with new commercial products. Because of the ambiguity of this new business, feedback to the EST/DE department continued to be slow and erratic.

Both of these problems impeded workers' attempts to develop an effective task structure. Under protected experimental conditions, workers would have been given the time and support. necessary to work through these problems; indeed, sufficient protection is needed to allow workers to develop, through a series of intermediate designs, a mature task structure. In this case, however, the demands to generate new commercial business did not afford workers the opportunity to develop experimentally a workable task design. Rather, they were forced to implement the new structure under full working conditions. This lack of experimental protection, when combined with inappropriate task skills and inadequate feedback, led workers to behave much as they had in the traditional departmental structure. Since the redesign disturbed these previous modes of behavior, workers experienced a great deal of anxiety in trying to perform in a task structure more unmanageable than the previous design.

Because of the problems inherent in implementing a redesign while trying to establish a new commercial market, the implementation process became untenable. Workers were frustrated with their inability to develop fully the redesign; this was especially evident in those instances where task demands intruded on the implementation process. Similarly, members from external departments experienced increasing difficulties in relating to the team structure, especially in the face of novel work requirements. Therefore, approximately six months after the start of the implementation stage, management decided to terminate the experiment and to return to the former departmental arrangement.

Before evaluating the experiment in terms of socio-technical theory and practice, it is important to note that the return to the former structure included some significant improvements derived from the experiment. First, a "producibility team," comprised of departmental supervisors and key numbers from sales and production, was placed at the input boundary of the department. This team screened inputs, set priorities, and assigned them to appropriate individuals within the department. This input function provided the department with needed task priorities as well as considerable protection from environmental forces; that is, it buffered the department's primary task from variable, external disruptions. Second, many of the environmental suggestions that were collected during the analysis were implemented. This included: clearer status reports, quicker estimating procedures, and a better way to track business through the company. Finally, workers showed considerable self-improvement as a result of their experimental involvement. For instance, they possessed greater knowledge of the EST/DE function; they devised new ways to perform some old tasks. They also had a clearer understanding of their environment, and they were better able to relate to each other around interdependent tasks. In summary, the experimental process, though not fully successful, led to some significant improvements.

EVALUATING THE EXPERIMENT

The EST/DE experiment provides a realistic portrayal of socio-technical experimentation. Although the value of learning from failure is often overrated, the results of the experiment are an important source of socio-technical learning. The experiment can be evaluated from three perspectives: (1) workers' perceptions of the redesign, (2) external departments' reactions to the redesign, and (3) consultants' views of the experimental process.

Workers Perceptions

The EST/DE redesign was an attempt to overcome many of the difficulties inherent in the previous departmental structure. As we have already discussed, the redesign was not implemented fully; this resulted in a good deal of frustration as workers were forced to perform within an undeveloped work structure. Here, we can examine some quantitative data to understand the impact of this restructure on workers' attitudes and behavior.

Research Design: To evaluate the effects of the redesign, a questionnaire was administered to the experimental group and to a control group at two points in time: one week prior to the implementation of the redesign and six months later. The comparison group was comprised of white collar workers from three departments in the company—personnel, engineering, and management information systems. A covariance analysis was used to examine posttest comparisons between the two groups. Using the pretest scores as covariants, it was expected that posttest differences would reflect the effects of the redesign. In other words, if the posttest scores were adjusted for pretest differences between the comparison groups, significant posttest differences would likely be the result of the experimental treatment.

Psychological Needs and Attitudes: The semantic differential instrument, developed for the socio-technical analysis, was used to assess the extent to which the redesign met workers' psychological needs and was evaluated as good or bad by workers. The nine bi-polar scales that were intended to measure Emery's (1963) psychological needs (see Figure 3) were factor analyzed into two need categories: task needs and esteem needs.

Table 6

PSYCHOLOGICAL NEEDS AND ATTITUDES: COVARIANCE ANALYSIS

	Pretest Mean	Posttest Adjusted Mean	F-Score Significance
Task Needs:			
Experimental group	3.2	4.5	F = 19.829
Control group	1.8	1.4	P = .001
Esteem Needs:			
Experimental group	6.0	6.5	F = 14.188
Control group	4.3	3.4	P = .001
Evaluative Attitudes:			
Experimental group	2.7	3.9	F = 5.602
Control group	2.2	1.8	P = .009

Note—Experimental group n = 22; control group n = 31.

The task needs factor was comprised of the following items: task variety, learning on-the-job, and setting standards of quality and/or quantity. The esteem needs factor consisted of three items: status and recognition, feedback of results, and recognition by others. In addition to psychological needs, a measure of workers' evaluative attitudes was also constructed using the SD items intended to measure this variable (see Figure 3): pleasant, valuable, and boring. Deficiency scores were computed for each of these measures, the difference between how much the department "should be" on a particular item and how much it "presently" was on the item. Since each deficiency score was the sum of three SD items measured on six interval, bi-polar scales, the scores ranged from 0 to 15. Table 6 lists the results of the covariance analysis for each deficiency measure: task needs, esteem needs, and evaluative attitudes. The data show that the experimental group has significantly higher deficiency scores on all three measures than the control group. This suggests that the departmental redesign resulted in a greater discrepancy between "what should be" and "what was" for task needs, esteem need, and evaluative attitudes.

The results of this analysis support the premise that the redesign provided fewer opportunities for meeting workers' psychological needs than the previous structure. Workers also evaluated the redesign as being worse than the previous design. Rather than attribute these negative reactions to the team structure, it seems more plausible that the failure to implement fully the redesign caused these perceptions. In other words, these data represent workers' frustrations of having to perform within an undeveloped task structure. The psychological impact of this situation is evident.

Work-Related Behavior. Table 7 reports questionnaire data about workers' performance: speed, cooperation, accuracy, and effectiveness. Workers were asked to rate their "present" performance and their "ideal" performance on 10-point scales. The scores in Table 7 are deficiency

Table 7

WORK-RELATED BEHAVIOR: COVARIANCE ANALYSIS

	Pretest Means	Posttest Adjusted Means	F-Score Significance
Speed:			
Experimental group	2.0	2.3	F = .566
Control group	2.3	2.0	P = .455
Cooperation:			
Experimental group	2.0	3.0	F = 8.711
Control group	2.3	1.4	P = .005
Accuracy:			
Experimental group	1.8	3.1	F = 7.060
Control group	2.1	2.0	P = .011
Effectiveness:			
Experimental group	2.5	3.1	F = 5.125
Control group	2.2	2.1	P = .028

Note—Experimental group n = 22; control group n = 31.

measures, the difference between the ideal and the present performance. The covariance analysis shows that the experimental group had significantly higher deficiency scores for cooperation, accuracy, and effectiveness than the control group. There is no significant difference between the comparison groups on the measure of speed.

The findings suggest that the redesign resulted in poorer performance. Workers experienced their cooperation, accuracy, and effectiveness as further from their ideal as a result of the redesign. These data are not surprising, since we would expect productivity to fall as workers attempt to implement and to adjust to the new work organization. Given the economic problems of the company, this reduced performance threatened the firm's survival in the commercial market. Thus, the decision to terminate the experiment appears to be a short-run solution to a pressing economic situation.

External Departments' Reactions

An additional source of information concerning the effects of the redesign is provided by the external departments interacting with the EST/DE department. During the implementation phase of the experiment, several members from these departments raised serious doubts about the appropriateness of the team structure. To better understand these external perceptions, quantitative data were collected.

The SD instrument used in the environmental analysis (see Figure 5) was administered to appropriate environmental members one week prior to the redesign and six months later. Respondents were asked to rate the EST/DE department on 23 bi-polar scales, with six intervals between the scales. Table 8 lists the responses to the SD instrument. Because the number of respondents differed between the two testing periods, parametric tests of significance on the pre and posttest differences were not computed. The data show that on 20 of the 23 scales, environmental members rated the EST/DE department as lower after the redesign than before it. A sign test reveals that this difference is highly significant (binomial distribution, $p = .002$, one-tailed test).

Table 8

EXTERNAL DEPARTMENTS' REACTIONS TO EST/DE DEPARTMENT

	Pretest (n = 39)		Posttest (n = 28)	
	Mean	S.D.	Mean	S.D.
Valuable	5.5	.555	4.8	1.228
Fast	4.1	1.142	3.8	1.206
Not wasteful of resources	4.3	1.019	3.7	1.362
Orderly	4.4	1.290	3.9	1.412
Essential	5.7	.521	5.1	1.331
Current	4.3	1.260	4.1	1.215
Reliable	4.8	.844	4.5	1.071
Potentially more important to my department	5.5	.644	5.3	1.090
Well regarded by others	4.5	1.097	3.9	1.585
Constructive	5.0	.811	4.5	1.071
Capable	5.2	.779	4.6	1.257
Becoming stronger	4.2	1.159	3.5	1.319
Important	5.6	.537	5.0	1.186
Accurate	4.4	.990	4.5	1.105
Easy to get along with	5.1	.990	4.7	1.182
On time with work needed by my department	4.1	1.450	3.8	1.624
Non-threatening	4.5	1.985	4.6	1.062
Does not need improvement in relations with my department	3.3	1.371	3.1	1.508
Open	4.9	.929	4.6	1.260

Table 8 *(Cont.)*

	Pretest (n = 39)		Posttest (n = 28)	
	Mean	S.D.	Mean	S.D.
Friendly	5.2	.970	4.6	1.133
Offers valuable advice to my department	4.6	1.482	4.6	1.129
Openly seeks advice and suggestions	4.3	1.555	3.9	1.474
Complete with information	4.8	.931	4.0	1.427

Note—For ease of interpretation, only the positive side of each bi-polar scale is reported.

An eye-ball examination of the larger decreases suggests that the EST/DE department was seen as less valuable, more wasteful of resources, less essential, less capable, becoming weaker, less important, less friendly, and less complete with information after the redesign than before it. These perceptions appear to be congruent with the workers' own views of the redesign. Again, these findings underscore the problem of trying to implement a new task organization under full working conditions. Given these negative environmental reactions, workers' feelings of low esteem and regard by others seem to be well grounded. Thus, rather than receiving support and protection for the experiment, workers were subject to negative environmental reactions.

Consultants' Views

The consultants saw the experiment as an invaluable opportunity to study and to apply the socio-technical approach in a white collar work setting. The failure to implement the redesign raises at least three questions:

1. Was the EST/DE department an appropriate experimental system?
2. Were the analysis and subsequent redesign faulty?
3. Was the company's economic crisis the major cause of the implementation failure?

A restrospective examination of the experiment may provide important clues to these questions. Let us review the experiment in terms of these issues. This should refine our understanding of the socio-technical approach, especially in regard to white collar work systems.

Appropriate Experimental System? The choice of experimental system determined, to a large extent, the course of the experiment. The initial interviews reveal that probability of success, chances for dissemination of results, and workers' interest in experimentation were adequate; however, two key characteristics of an appropriate experimental system were questionable: (1) a clearly differentiated work system and (2) clear measures of inputs and outputs. Since a clearly differentiated work system provides the task coherence required for analysis and redesign, a faulty judgment at this stage of experimentation is critical. The criteria for bounding a work system—time, territory, technology, and psychological and sociological attributes—did not produce a clear solution to the boundary problem. Because of certain task interrelationships between the EST and DE departments, it was decided to combine them into a single work system. The fact that the units shared specific time, territory, and sociological attributes supported this choice.

The decision to form an EST/DE department appears questionable in light of three problems that were relatively overlooked during the preliminary analysis. First, EST and DE workers identified psychologically with their respective primary tasks. This tended to place a mental barrier between the EST and DE functions, hence reducing the integration needed for system connectedness. Second, the technologies and skills of each function were so different that task segregation within the department was difficult to overcome. Finally, the task interrelationships between the EST and DE functions were relatively weak in comparison to task relationships across the boundary of the department. For instance, the EST unit, which provided inputs to larger sales decisions, was connected more to the sales department than to the DE unit; similarly, the DE function, which provided inputs to the die shop, was related more to this department than to the EST unit.

These problems raise serious doubts about the boundary of the experimental system. Furthermore, they suggest that the boundary criteria must be used with a great deal of caution and sensitiveness if an appropriate boundary is to be enacted. Failure to realize that technological and psychological attributes were more important in bounding the system than the criteria of time, territory, and sociological attributes resulted in an internally segregated experimental system. Also, failure to weight the importance and strength of task interrelationships within and without the department resulted in a work system the components of which were related more to external components than to each other. In retrospect, these oversights produced an unclearly differentiated experimental system, and this had a significant impact on the experimental analysis and redesign.

The second problem in choosing an appropriate experimental system, clear measures of inputs and outputs, was not resolved during the experiment. The preliminary interviews showed this to be a difficulty, yet it was decided to proceed with the experiment under the assumption that the analysis would produce clear measures of these variables. Unfortunately, the analysis explored the consequences of this problem rather than solve it. It was shown that the EST/DE department's inputs and outputs were both variable and unique; without clear task priorities and standards for assessing quality and utility of performance, workers were unable to balance their work loads or to know how well or poorly they were doing. Instead of devising a procedure for measuring and placing priorities on inputs and outputs, the consultants, faced with severe time pressures, took the more expedient alternative of postponing this task until the redesign phase of the experiment. This proved to be a major mistake as it was difficult to assess the various redesign hypotheses without clear criteria for system effectiveness. Indeed, the lack of clear measures of inputs and outputs made it almost impossible to understand the rationality of workers' behavior and to design a more rationally functioning system. In review, the decision to experiment with a work system with few measures of inputs and outputs appears questionable. A more effective strategy would have been to solve this problem prior to the experiment or, if this was not possible, to choose a more appropriate experimental system.

In summary, two criteria for an appropriate experimental system were relatively unsatisfied in this study: a clearly differentiated work system and clear measures of inputs and outputs. In the absence of these characteristics, the experimental unit did not have the necessary coherence or standards of performance to be analyzed and redesigned properly. Both the boundary decision and the choice to experiment with a unit with few measures of inputs and outputs were subjective judgments. Failure to use the boundary criteria with sufficient sensitiveness and the pressure to proceed with the experiment resulted in faulty judgments on both of these issues. Although we cannot judge the importance of the other conditions for effective experimentation—probability of success, chances for dissemination, and workers' interest—they do not appear to ensure success in the absence of these characteristics.

Faulty Analysis and Redesign? The socio-technical analysis provided a systematic critique of the EST/DE department; however, an important part of the data was neglected. This concerns the differences between the EST and DE functions. As we have already discussed, the decision to bound the separate functions into a single work system influenced the subsequent stages of the experiment. In regard to the analysis, the boundary decision biased the interpretation of the data. Each step of the analysis revealed significant differences between the EST and DE functions—e.g., objectives, role structures, psychological needs, environmental interfaces, and environmental perceptions. Rather than see these data as a disconfirmation of the experimental boundary, they were viewed in reference to a presumed, internally-coherent system. This conception influenced the subsequent hypotheses for redesign. It tended to limit the hypothesized redesigns to those integrating the EST and DE functions into common work units. In retrospect, such designs may not have been feasible or appropriate, given the technological and psychological differences between EST and DE. Perhaps, had the differences inherent in the data been recognized, more relevant redesigns might have been hypothesized.

In review, the analysis and hypotheses for redesign failed to account for significant differences between the EST and DE functions. Failure to see these differences may be attributed to the assumption that the work system was internally coherent. This suggests that there is a powerful impetus to see things in terms of the initial definition of the situation. Perhaps some form of cognitive dissonance contributes to this biased interpretation of subsequent data (Festinger, 1957). If so, this points to a potential difficulty in socio-technical experimentation: the tendency to defend, often implicitly, the initial boundary definition of the work system. Given the arbitrary nature of this boundary, there is a good probability that it may be wrong. It is critical, therefore, to realize that the initial boundary is only a subjective judgment, open to further testing during the analytical phase of the experiment.

Cause of Implementation Failure? The failure to implement the redesign was attributed to the economic crisis of the company—that is, the need to generate new commercial business placed a larger work demand on the EST/DE department, hence reducing considerably the protection required for experimentation. Given the need for substantial exemption from normal operating demands during the implementation phase, it is not surprising that workers were unable to implement fully the redesign. Indeed, the anxiety aroused by this untenable situation precluded much experimentation. While the impact of the firm's economic crisis cannot be denied, two additional factors may have contributed to the implementation failure: the specification of the redesign and resistance to change.

The specification of the redesign followed from the process of developmental system design outlined in Chapter 6. The self-regulating conditions necessary for autonomous group functioning were specified, and the remaining dimension was left free to develop in accordance with the needs of the work groups. Theoretically, this amount of clarity seemed appropriate for the departmental redesign. It was not suspected, however, that the new design might be underspecified, especially in light of its radical departure from the existing structure. This is a pertinent point, for workers complained that the team structure was too ambiguous. It left too many task-relevant issues vague, such as task relationships, responsibilities, boundary interfaces, and load distributions. In the absence of a well-developed decision-making process and experience in team forms of production, workers were unable to function as a work group.

Autonomous group structures seem particularly susceptible to the problem of underspecification of design; especially in those situations where the design is a radical departure from the existing task structure or where the conditions for self-regulation are not met fully. In this study, both of these problems were present. Failure to realize their significance resulted in a redesign beyond workers' capacities to operate. Had the process of developmental system design been followed properly, we might have started with a relatively detailed structure and proceeded to less specified designs as workers became more competent in this form of production. Since this process proceeds through intermediate work structures to a more fully developed design, it accounts for the dynamics of social change which involve much learning and experimentation as new work structures are developed.

Our conceptualization of workers' resistance to change concerns an active seeking toward positive affirmation: When redesigns disrupt existing sources of affirmation, workers will seek to preserve those structures that provide such confirmation. In many ways the team redesign negated several sources of existing affirmation—e.g., functional skill identity, individual role identity, and the security associated with a well-defined job. In the absence of these conditions, several workers sought to preserve their old departmental roles and relationships. Put differently, these individuals resisted many of the work team changes. Although much of this activity was relatively covert, such as maintaining departmental relationships no longer needed in the team structure, it served to preserve the positive feelings disrupted by the redesign. Given the necessary experimental protection, it was expected that the redesign would gradually provide new sources of confirmation—e.g., team identity, challenging tasks, and better environmental relationships. Since such protection was lacking, a more thorough examination of this problem might have uncovered other ways to provide confirmation. Perhaps a more structured redesign which preserved some of the existing conditions would have resolved it.

Summary. A retrospective examination of the experiment provides important clues about the results of the experiment. This refines our understanding of the socio-technical approach and improves further application of this strategy. The choice of an appropriate experimental system appears questionable in this study. The lack of a clearly differentiated work system and few measures of inputs and outputs resulted in an experimental system that was neither internally coherent nor rationally defined. This choice had significant consequences on the analytical and redesign phases of the experiment. It resulted in a biased analysis that neglected significant differences between the EST and DE functions; this, in turn, led to a redesign that may have been too integrative, given the differences between the functional units. Finally, the company's economic crisis, when combined with an underspecified redesign with few opportunities for positive affirmation, contributed to the implementation failure.

CONCLUSION

The estimating and die engineering experiment is a realistic illustration of socio-technical experimentation. The theory and strategy of socio-technical systems provide a sound basis for such experiments; yet, the conditions that emerge in the actual work situation largely determine the course this approach will take. We have sought to portray the experiment as it actually happened. Though our reconstructed logic makes the process of experimentation appear more orderly than it is, we have tried to present some flavor for the judgments involved in work system change.

The following chapter describes a wheel-line experiment in the same company. This experiment, in a blue collar setting, is especially interesting, since it was a major attempt to resolve the company's economic crisis by expanding its commercial business.

9

THE WHEEL-LINE EXPERIMENT: A CASE STUDY OF BLUE COLLAR WORK DESIGN

The wheel-line experiment is an application of the socio-technical approach to a blue collar work setting. In many ways, this study is a sequel to the EST/DE experiment described in Chapter 8. Whereas that experiment was halted abruptly by the company's economic crisis, the wheel line was a significant effort to resolve this crisis. Temporally, the two experiments over-lapped considerably, since much of the work load encountered during the EST/DE study involved the preparatory tasks required to obtain and to start up the wheel business. Somehow, it seems ironic that the work demands which thwarted the EST/DE experiment caused the start of the wheel-line experiment. Again, this points to the powerful impact of situational factors on the experimental process, and many of the organizational forces affecting the EST/DE department had a similar impact on the wheel line.

BACKGROUND

The specific circumstances leading to the wheel-line experiment involved a rapid shift in the market for the firm's products. For a number of years prior to the experiment, the company's products were oriented primarily to an aerospace market with a small percentage of forgings aimed at a commercial market. One of the commercial products was aluminum truck and bus wheels which were forged in-house, sent to subcontractors for finishing, and shipped back to the company for delivery to customers. This arrangement allowed the company to concentrate its resources in forging and sales without having to invest directly in costly finishing equipment.

In the late 1960s the aerospace market began to decline rapidly. To counter the loss of aerospace business, the firm sought ways to expand its commercial market, and an outcome of this search was a substantial contract to forge automobile wheels for a major automotive company. A primary requirement of the contract was the continuous delivery of a large volume of wheels directly to the automobile assembly lines. Since this would require coordination and control over all facets of production as well as a supply of wheels greater than that which could be processed by the subcontractors, top management decided to purchase finishing equipment from one of its subcontractors, thus creating an integrated wheel business.

The start-up of the new finishing operation presented two major problems. The first was concerned with the design of the finishing jobs. The finishing tasks involved a sequential series of machining, drilling, and packaging activities; both union and management officials felt that the company's employees would not be attracted to these jobs if they were designed traditionally into assembly-line jobs. This argument was based on the fact that the finishing jobs would be filled from a pool of union workers who were accustomed to the more challenging kinds of jobs associated with a forging or job-shop technology. The second problem was time: the company had to install equipment, train workers, and make the transition to normal operating conditions within a few months. This placed a substantial strain on the firm's ability to cope with the obstacles commonly encountered in new production start-ups—e.g., resistance to change and balancing the production line. Since both problems, the design of jobs and time constraints, were related to the company's ability to compete in the commercial market, the union and management decided to sanction jointly a socio-technical experiment to help the company survive in this new market.

The wheel-line experiment started in November 1971 and lasted for about one year. The specific events of this period included: (1) identifying the experimental system, (2) sanctioning the experiment, (3) analyzing the wheel line, (4) generating hypotheses for redesign, (5) training workers, (6) implementing the redesign, and (7) evaluating the experiment.

IDENTIFYING THE EXPERIMENTAL SYSTEM

The finishing operation appeared to be an appropriate experimental system. Its machining, drilling, and packaging tasks were interdependent and formed a relatively self-completing whole; its inputs and outputs were clearly defined and easy to measure. Chances for success were judged good, since the unit would be a new start-up, free from many traditional constraints and supported by both the union and management. Opportunities for dissemination of results seemed adequate, especially in the context of other plants in the company with similar long-linked technologies. Finally, workers' interest in experimentation was difficult to assess because the finishing operation was not in operation at the firm. Preliminary talks with the union suggested that workers would pose few problems. Based on these judgments, the finishing operation appeared to be an almost ideal experimental system. Furthermore, it was suggested that if the wheel business increased, the boundary of the finishing operation might be expanded to include relevant forging and sales functions, thus forming a fully integrated wheel system.

SANCTIONING THE EXPERIMENT

The primary objective of the sanctioning process was to attain the necessary commitment and protection from top management and local union officials to engage in socio-technical experimentation. Starting with union and management's concern for the success of the wheel business, the sanctioning process involved a series of union/management meetings in which the initial design of the finishing jobs and rules for the experiment were worked out. The initial stages of this process involved an introduction to socio-technical concepts and a mutual testing out between the consulting team—consisting of an external and internal consultant—and the union representatives. This was followed by a socio-technical analysis of the finishing operation and a tentative design for the finishing jobs.

After the analysis and design activities, the sanctioning process involved a series of agreements concerning wage rates, protection of seniority, bidding order for the new jobs, the status of the finishing operations in the existing collective bargaining contract, and the location of the finishing operations in the representation structure of the union. Although it was preferable to resolve these issues apart from the normal collective bargaining process, these latter activities involved heated debates, closed door caucuses, and informal bargaining sessions between union and management officials. The consultants' primary role was to clarify issues and facilitate communication between the involved parties.

Since the union and management were already operating under a negotiated contract, the experiment was equated to the start-up of a new department. Under existing agreements, normal methods of job evaluation were suspended temporarily, providing workers with the opportunity to experiment with new wheel-line designs. Additional sanctioning rules included:

1. The implementation phase of the experiment would run for four months.
2. Wheel-line jobs would be posted, and bidding would be based on total company seniority.
3. Up to 30 days, workers could leave the experiment and return to their prior jobs.
4. Workers would have 30 days to meet minimum job requirements as specified in the original posting.
5. Workers would retain company-wide seniority and union rights.
6. Posted wages would be frozen during the length of the implementation phase. A union/management evaluation team would reevaluate the jobs after this phase.
7. Normal levels of productivity would be suspended during the implementation phase, and a union/management committee from the wheel line would decide on standard levels of productivity after this phase.
8. The implementation phase would take place within the budgetary and manpower constraints of the company.

ANALYZING THE WHEEL LINE

Since the wheel line was to be a new operation for the company, two unusual conditions existed at this stage of the experiment. First, the union demanded full participation in the analysis and hypotheses generation phases before they would sanction the experiment. Thus, these stages of the experiment were conducted as part of the sanctioning process, although we present them separately for ease of understanding. Second, at this time, the equipment purchased from the subcontractor was still in operation at his plant; therefore, it was necessary to do a preliminary analysis of the wheel line as it existed at the subcontractor's plant. This allowed for redesign proposals based on its existing operation, but with little knowledge about how it would operate at the company. To overcome this problem, the analysis and hypotheses generation phases of the experiment were considered tentative, with the expectation that further analysis and hypotheses generation would probably have to be carried out during the implementation stage of the experiment.

Two managers from the company, the union sanctioning committee, and the consultants carried out the analysis. Several trips were made to the subcontractor's plant to observe the wheel line in operation and to collect various kinds of data. The analysis was guided by an analytical model presented in detail in the Appendix. The major steps included: analyses of the technical system, the social system, and the environment.

Analysis of the Technical System

The technical analysis included an initial scanning of the work system to determine its inputs, transformations, and outputs; the geographical layout of its operations was also examined. Against this background, the unit operators or main phases of the production process were located, and key process variances were identified. These data provided initial clues as to the technical problems of the unit and set the stage for the analysis of the social system.

Initial Scanning. The primary input into the wheel line is forged wheels. Conversion is finishing the wheels and packaging them for sale. Output is packaged wheels delivered to the shipping department of the X Company for subsequent delivery to customers.

The operations for converting a forged wheel into a finished product consist of a series of machining, pre-stressing, drilling, cleaning, and packaging processes. The operations are linked sequentially so that the output of the first operation serves as the input to the second and so on. The geographical layout and conversions of the wheel line are represented graphically in Figure 1.

The outside and inside diameters of a forged wheel are bored and finished on automatic, milling machines. Both diameters are then pre-stressed on an automatic, pre-stressor. Bolt holes, hand holes, and valve slots are drilled on automatic vertical drills, and then deburred with a hand-held, electric deburrer. Hand holes and valve slots are pre-stressed on a vertical, pneumatic pre-stressor, and the bolt holes are counter-sunk on a manually controlled counter-sinker. Identification numbers are hand stamped into each wheel, and final deburring completed. Approximately 25 percent of the wheels are then polished and buffed on a belt-driven

Figure 1

GEOGRAPHICAL LAYOUT AND OPERATIONS OF THE WHEEL LINE

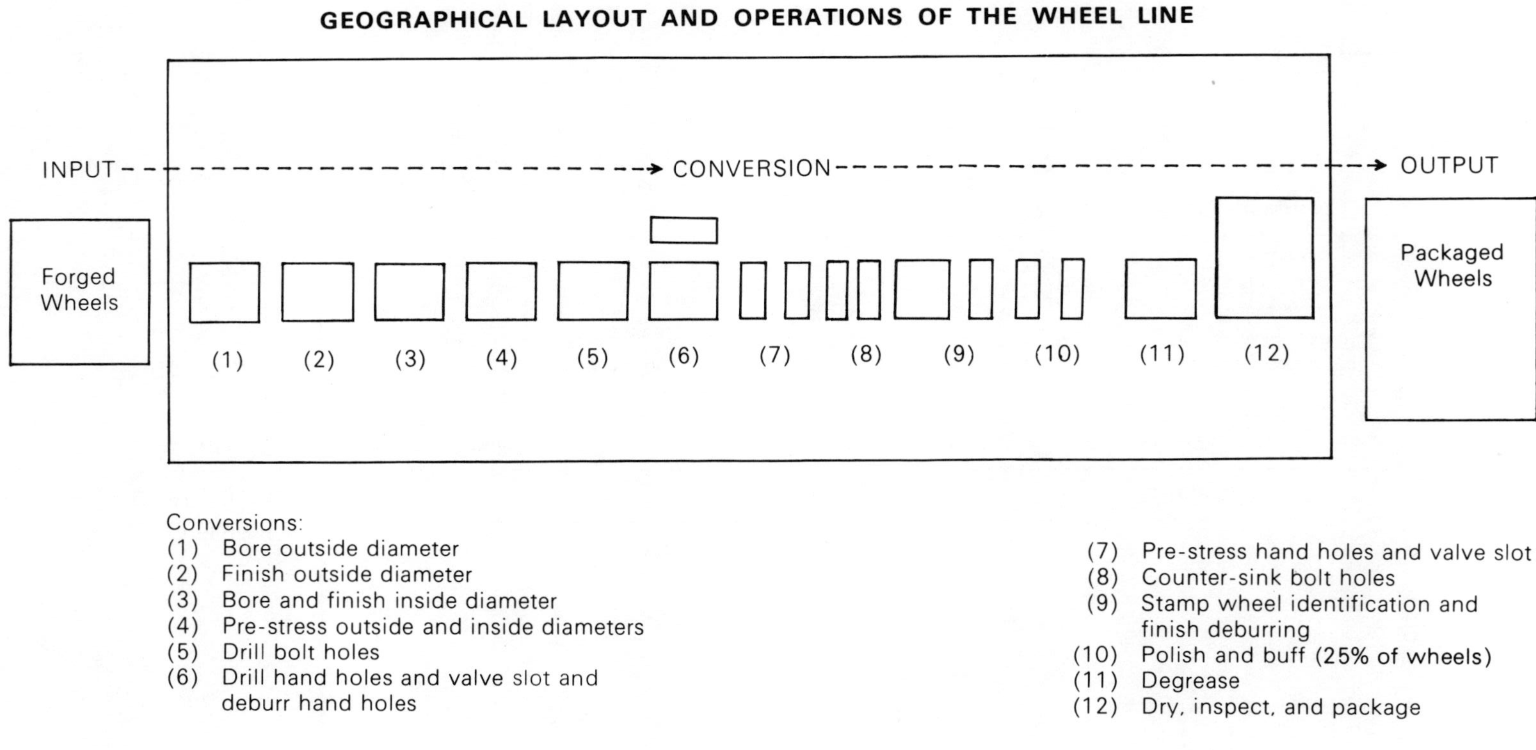

Conversions:

(1) Bore outside diameter
(2) Finish outside diameter
(3) Bore and finish inside diameter
(4) Pre-stress outside and inside diameters
(5) Drill bolt holes
(6) Drill hand holes and valve slot and deburr hand holes
(7) Pre-stress hand holes and valve slot
(8) Counter-sink bolt holes
(9) Stamp wheel identification and finish deburring
(10) Polish and buff (25% of wheels)
(11) Degrease
(12) Dry, inspect, and package

Table 1

UNIT OPERATIONS

Unit Operations	Inputs	Transformations	Outputs
MACHINING	Forged wheels; machining coolant	Bore, finish, and pre-stress outside and inside diameters	Finished outside and inside diameter wheels
DRILLING	Finished outside and inside diameter wheels; drilling lubricant	Drill bolt holes, hand holes and valve slot; deburr hand holes; pre-stress hand holes and valve slot; counter-sink bolt holes	Finished bolt holes, hand holes and valve slot wheels
FINISHING	Finished bolt holes, hand holes, and valve slot wheels; identification numbers; packaging materials; degreasing solvent; polishing lubricant	Stamp identification; final deburr; polish and buff; degrease; dry; inspect; package	Packaged wheels

polisher and buffer. Finally, all wheels are degreased in a large tank of solvents, hand dried, visually inspected, and packaged.

The line is in operation for one, eight-hour shift per day, and about 110 wheels are finished per shift. The activities for each conversion process are cyclic—load, operate, unload. Since operating time greatly exceeds loading and unloading time, operators of automatic machines clear away excess scrap, monitor the machine, and prepare for the next cycle of operations when the machine is running.

Unit Operations. The wheel line's conversion process may be broken down into three main phases or unit operations—boring, drilling, finishing. Each unit operation is a relatively self-contained segment of the conversion process with an identifiable set of inputs, transformations, and outputs. Table 1 presents a detailed description of each unit operation.

Process Variances. The process variances of the conversion process were examined to identify major control problems. Variance is taken to mean a deviation from some standard or specification, and we were concerned only with those variances that arise from the raw material or from the nature of the production process. Table 2 is a matrix of the process variances of the wheel line. Those variances that significantly affect quantity or quality of production—or operating costs or social costs—are considered key. The location of each variance is specified in relation to the unit operations so that localized clusters of variance may be identified.

A review of the variance matrix reveals the following variance control problems: (1) boring operation—the amount of flash, type of alloy, and machine cycle speed affect the state of the tooling; (2) drilling operation—the type of alloy and machine cycle speed affect the state of the drill bit; and (3) finishing operation—the type of alloy and machine cycle speed affect the state of the polishing abrasive. Since the amount of flash, type of alloy, machine cycle speed, state of tooling, and state of drill bit affect the finish of the wheel, the quantity of wheels produced, and also the costs of tooling and drill bits, they are considered key.

Summary of Problems Associated with Technical System

1. The interdependence of the tasks makes the cycle of operations susceptible to disruption; thus slowing down or speeding up any single operation affects all other operations.
2. The key process variances affect operations in each unit operation, thus increasing the likelihood of disruption of the overall cycle of operations. The type of alloy and machine cycle speed affect the state of machine tooling (boring unit), the state of drill bits (drilling unit), and the state of the polishing abrasive (finishing unit).
3. The amount of flash on the forged wheels creates an excessive waste removal problem in the boring unit, thus making this area extremely dirty and potentially dangerous.

Analysis of Social System

The social analysis examined the work roles in the unit—their tasks,

Table 2

MATRIX OF VARIANCES

Direction of Variance →	Amount of Flash	Type of Alloy	Machine Cycle Speed	Machine Feed Speed	State of Tooling	State of Drill Bit	State of Polishing Abrasive	State of Degreasing Solvent	Unit Operations
Amount of Flash	(K)								BORING
Type of Alloy		(K)							
Machine Cycle Speed			(K)						
Machine Feed Speed				○					
State of Tooling	○	○	○		(K)				
State of Drill Bit		○	○			(K)			DRILLING
State of Polishing Abrasive		○	○				○		FINISHING
State of Degreasing Solvent								○	

Note: (K) = Key Variances.

Table 3

OCCUPATIONAL ROLES

Occupational Role	Number/ Role	Location	Tasks
Supervisor	1	Boring, drilling, and finishing	Schedule work flow and major maintenance; regulate and control workers; trouble shoot; schedule input and output inventory; account to top management for quality and quantity of output.
Machinist	4	Boring	Manually place wheel in machine; monitor machine operation; remove waste from machine; inspect wheel; hand wheel to next operation; minor machine maintenance; inspect and change tooling.
Drill Operator	4	Drilling	Manually place wheel in machine; monitor drill operation; remove waste from drill; inspect wheel; hand wheel to next operation; minor drill maintenance; inspect and change drill bits.
Polisher	1	Finishing	Manually place wheel in machine; monitor polishing operation; remove waste from polisher; inspect wheel; hand wheel to next operation; minor maintenance; inspect and change abrasives.
Finishing Helper	4	Boring, drilling	Stamp and deburr wheel; manually place wheel into and out of degreaser; inspect and change solvent; clean wheel; final inspection; packaging; act as roving inspector for wheel line.
Utility Man	1	Boring, drilling, and finishing	Keep production area clean; bring coolants and solvents when needed.

Table 4

ANALYSIS OF VARIANCE CONTROL

Key Process Variance	Name of Unit Operation: Where Occurs	Where Observed	Where Controlled	By Whom	Control Activities	Information Related to Control Activities	Hypotheses for Re-design
Amount of Flash	Boring	Boring	X Company	Forger	Adjust press and die	Amount of excess flash over specification	Inspect incoming forged wheels for undue flash
Type of Alloy	Boring, drilling and finishing	Boring, drilling and finishing	X Company	Forger	—	Type of alloy vs. which use	Schedule optimal runs of one alloy at a time
Machine Cycle Speed	Boring, drilling and finishing	Boring, drilling and finishing	Boring, drilling and finishing	Machinist, drill operator, polisher	Adjust RPM by minor gear change	RPM vs. state of wheel	Variable speed control set by operators within tolerance limits
State of Tooling	Boring	Boring	Boring	Machinist	Change tooling	State of tooling vs. state of wheel	Schedule tool changes for all machines
State of Drill Bit	Drilling	Drilling	Drilling	Drill operator	Change drill bit	State of drill bit vs. state of wheel	Schedule drill bit changes for drills

their geographical location, and their interrelationships. Then, the key process variances were analyzed in terms of variance control. Finally, the roles in the work system were evaluated in reference to psychological needs.

Role Analysis. Fifteen workers carry out the operations of the wheel line. They are divided into six occupational roles according to skill level (by decreasing level of skill): (1) supervisor, (2) machinist, (3) drill operator, (4) polisher, (5) finishing helper, and (6) utility man. Table 3 lists the occupational roles, number of workers per role, unit operation location, and tasks of the role holders.

The geographical relationship among the occupational roles during a normal eight-hour shift is quite stable. The supervisor spends about one-fourth of his time observing the various unit operations by walking around the shop floor. The remainder of his day is spent either in his office located adjacent to the wheel line or in other parts of the plant. The machinists, drill operators, and polisher are permanently assigned to a particular machine. The machine cycles allow for minimal conversation with those on either side of the machine and walking about is reduced to a small area directly in front of the machine. The finishing helpers rotate among the various operations in the finishing unit about twice a week; they also perform a roving inspection task in the boring and drilling operations. The utility man cleans the entire working area, but because more waste is accumulated in the boring unit, about one-half of his time is spent working among the first three machining operations.

Variance Control. One of the primary responsibilities of the social system is to control the key process variances identified in the previous analysis of the technical system. To discover the extent to which the key variances are presently controlled by the social system, a table of variance control was constructed. This allowed for identification of variance control problems as well as suggestions for change. Table 4 presents the analysis of key variance control.

The table of variance control reveals that two key variances are outside the control of the wheel line: the amount of flash and the type of alloy. Both of these variances are imported into the wheel line from the wheel-forging operations of the X Company, and are thus under the control of the company. Feedback from the wheel line to the X Company regarding these variances is extremely slow, causing some disruption on the wheel line before corrective action is taken. Inspection of incoming wheels and scheduling around the type of alloy could help to alleviate these problems.

The remaining variances, machine cycle speed, state of tooling, and state of drill bit, occur, are observed, and are controlled on the wheel line. Although the feedback cycle is relatively short in contrast to the externally imported variances, corrective action varies as to length of shutdown time required. Changing machine cycle speeds is extremely time consuming and is constrained by the machine-speed requirements for finished wheels. Changing tooling and drill bits is relatively less time consuming, but differential rates of wear across the machines make this process disruptive for the overall cycles of machining and drilling. Shutting down one machine for a tool change, for example, disrupts the flow of materials to and from the other machining and drilling operations. Designing the machines for variable speed control could help to eliminate the long shutdown times and might help to balance the overall cycle of machining and drilling operations. Scheduling tooling and drill bit changes for the entire line in advance might help to alleviate the disruptive character of present shutdowns.

Psychological Needs. Based upon evidence that people have work requirements beyond those usually specified in a contract of employment, a general set of psychological needs pertaining to the content of jobs has been developed (Emery, 1963). Table 5 lists the psychological needs and examines each role against the list on a simple adequate/inadequate basis. Since it was not possible to interview the role holders as to their perceptions of their jobs, the data are based upon external observations and a good deal of empathetic guessing.

One of the most striking findings of the psychological needs analysis is the inadequacy of present job designs for meeting basic psychological needs. This is especially evident among the machinists, drill operators, and polisher whose jobs appear adequate on only one psychological dimension—"amount of decision-making." Although the finishing helpers' and utility man's jobs appear adequate on more of the psychological dimensions, this can be attributed primarily to the wider variety in their jobs and to their mobility along all parts of the wheel line.

Summary of Problems of Social System

1. The assignment of machinists, drill operators, and the polisher to one operation engenders a limited orientation to the overall cycle of operations and leads to minimal social contact.
2. The machinists in the boring unit are dependent on the utility man to remove excess waste before it accumulates. This leads to increased anxiety and conflict as the most skilled role holders (machinists) are dependent upon the least skilled role holder (utility man). Since excess chips make this work area dirty and potentially dangerous, interpersonal conflict is potentially magnified.
3. Two key process variances—amount of flash and type of alloy—are outside of the control of the operators of the wheel line. Since these variances affect quality and quantity of production as well as tooling and drill bit costs, their control outside the unit is not adequate for system regulation.
4. Control of machine-cycle speeds is difficult to influence as the cycles are pre-set according to machine tolerances and the wheel finish required.
5. The state of tooling, drill bits, and polishing abrasive depend on three factors—amount of flash, type of alloy, and machine cycle speed—not directly under the control of operators of the wheel line. Scheduling for tooling and drill bit changes is difficult and places the workers in a stressful position in

Table 5

PSYCHOLOGICAL NEEDS

Psychological Needs	(ROLES) Machinist	Operator	Polisher	Helper	Utility Man
The need for being able to learn on the job and go on learning	Inadequate	Inadequate	Inadequate	Inadequate	Inadequate
The need for the content of the job to be reasonably demanding in terms other than sheer endurance and yet providing a minimum of variety (not necessarily novelty).	Inadequate	Inadequate	Inadequate	Adequate	Adequate
The need for some minimal amount of decision making that the individual can call his own.	Adequate	Adequate	Adequate	Adequate	Adequate
The need for some minimal degree of social support and recognition in the workplace.	Inadequate	Inadequate	Inadequate	Adequate	Inadequate
The need to be able to relate what he does and what he produces to his social life.	Inadequate	Inadequate	Inadequate	Inadequate	Inadequate
The need to feel that the job leads to some sort of desirable future.	Inadequate	Inadequate	Inadequate	Inadequate	Inadequate

regard to other operations when unscheduled changes disrupt the overall flow of operations.

6. The present design of jobs does not meet minimal psychological needs. Present jobs are relatively repetitive, boring, and do not provide for learning or social recognition. This is especially true for the machinists, drill operators, and the polisher who follow a one-man/one-machine design.

Analysis of Environment

To complete the socio-technical analysis, important environmental interfaces were examined. Since critical variances were imported into the socio-technical system from environmental units, consideration of these potential problem areas was essential for the redesign proposals. The supply and user systems as well as the maintenance system were considered in this context.

Supply System. The supply system for the wheel line is the wheel-forging operation of the X Company. Forged wheels are stacked 36 to a pallet and sent by truck to the wheel line. A fort-lift truck unloads the truck and places the pallets in a storage area adjacent to the boring unit. Approximately every 2½ hours, a full pallet is placed before the first machining operation.

Two important variances are introduced into the wheel line from the forging works of the X Company—the amount of flash on each wheel and the type of alloy. Both of these have been discussed with the conclusion that both remain outside of the control of the wheel-line operators.

An additional problem introduced into the wheel line by the forging works is the placement of the forged wheels on the pallet. Wheels are placed face down on the pallet, thus causing the first machine operator to perform an extra lifting operation as he must turn each wheel over before placing it in the machine. Since the wheels weigh about 90 pounds, this extra physical exertion causes undue fatigue over an eight-hour shift.

User System. The user system for the wheel line is the sales department of the X Company. Finished wheels are stacked 36 to a pallet and trucked back to the X Company for delivery to customers. Present quality control and production standards are adequate, but customer complaints are routed through sales to the management of the wheel line. This results in a long feedback cycle before corrective action can be taken on the wheel line. Wheel-line operators must often wait months to receive feedback as to the results of their work, and often this arrives in the form of nebulous complaints not easily traced to the source of the difficulty.

Maintenance System. Minor maintenance such as machine and tooling adjustments are carried out by the machinists, drill operators, and the polisher. Additional maintenance is done by outside contractors available on an emergency-call basis. Although little preventative maintenance is carried out, major maintenance problems have not been crucial to wheel-line operations. One explanation for this lack of maintenance problems may be the relative simplicity of the machinery in addition to the single-shift of operations.

Summary of Environmental Problems

1. The supply system introduces two main variances into the wheel line—the amount of flash and the type of alloy. Both variances affect output and costs and are relatively outside of the control of the wheel line.
2. Placement of wheels on pallets affects the amount of physical lifting carried out on the first boring operation.
3. Feedback from customers to the wheel line is quite slow and is routed through the sales department of the X Company.
4. A lack of preventative maintenance could result in a more costly breakdown in the long-run.

GENERATING HYPOTHESES FOR REDESIGN

Data from the preliminary analysis of the wheel line as it existed at the subcontractor's plant were used to tentatively redesign the line. Instead of presenting all possible proposals for change, only those that were experimented with at the X Company are described.

A Group of Workers for a Set of Interdependent Tasks

A major design criterion at the group level is to "provide for interlocking tasks, job rotation or physical proximity where there is a necessary interdependence of jobs" (Emery, 1963). This may help to sustain communication and to create mutual understanding between workers whose tasks are interdependent. The wheel line included two sets of relatively interdependent tasks: machining and drilling operations in the first set, and stamping, polishing, degreasing, and packaging operations in the second set. Two autonomous work groups were created for each set of interdependent operations. Each group was to contain workers with the requisite skills for carrying out all of the operations in each set, and the two groups were to be physically separated by a buffer inventory of approximately 120 wheels (enough for one shift of operations).

Each work group was to be responsible for carrying out all of the operations in its particular set, but with minimal specifications as to the specifics of task design. This would allow for detailed job designs to evolve depending upon the individual needs of the workers and the particular task requirements of the various operations. The redesign would also provide each group with some semblance of an overall task.

If the groups were to function autonomously, provisions for setting standards, receiving feedback of results, and controlling boundary conditions were essential. To fulfill these requirements, each group was to schedule daily production within the constraints of a weekly production quota set by management. Since both groups were relatively interdependent—except for the 120-wheel inventory separating them—some degree of inter-group scheduling was needed. Again, this was to be left to members from each group. Feedback of results was to be a daily production record which stated the number of acceptable wheels produced by both groups as well as changes in the amount of wheels in inventory. Templates would provide direct feedback for each machining operation.

Providing some control over boundary operations would be a difficult task. Although input and output inventories would help to buffer each group from their supply and user systems, some form of communication system was essential for reducing some of the variances exported into each group by the external departments. Members of each group were to be placed into formal contact with members from each of the interfacing departments. Meetings were to be scheduled between these groupings so that emergent problems could be met within the context of on-going work relationships. For example, the machining and drilling group would meet with the wheel-forging department, while the polishing, stamping, degreasing, and packaging group would meet with the sales department. Table 6 summarizes the roles and tasks of the redesign.

Although the autonomous group redesign could have been specified more fully, it was felt that each group could better regulate its activities if many of the design variables were left free to vary with the situation. This would allow each group some freedom as to task design and work relationships in addition to providing some degree of maneuverability during the implementation phase of the experiment.

The Manager as Boundary Maintainer

The redesign of the wheel line into two autonomous work groups

Table 6

OCCUPATIONAL ROLES, NUMBER PER ROLE, AND TASKS OF WHEEL LINE REDESIGN

Occupational Role	Number	Tasks
Supervisor	1	Help maintain departmental interfaces; account to top management for quantity and quality of output; planning; help workers with social and technical problems if needed.
Group A Machining and Drilling	6	Boring and drilling operations; minor maintenance; schedule input and output inventories; interface with supply, user, and maintenance systems; interface with Group B; inspect wheels; clean area, maintain and change tooling, drill bits, and machine libricants; interface with supervisor.
Group B Stamping, Polishing, Degreasing, Packaging	6	Stamping, polishing, degreasing, and packaging operations; minor maintenance; schedule input and output inventories; interface with supply, user and maintenance systems; interface with Group A; maintain and change polishing abrasives and degreasing solvents; final inspection; interface with supervisor.

suggested a new role for the foreman. Since it was expected that each group would be relatively self-regulating, the foreman would not have to spend as much time controlling internal work relationships. Instead, his primary duties would be relating the wheel line to the other departments in the company as well as planning for the future of the wheel line. In addition, he would help both groups solve social and technical problems.

Viewed in this manner, the foreman would work with the members of each group to help them schedule production, receive feedback of results, maintain some control over environmental interfaces, and solve social and technical problems. He would give the wheel line some constancy of direction through planning while providing overall boundaries of discretion within which to operate.

SUMMARY OF SOCIO-TECHNICAL ANALYSIS AND REDESIGN PROPOSALS

The purpose of these last two stages of the experiment has been to analyze the wheel line as it existed at the subcontractor's plant, and to provide redesign proposals for possible implementation at the X Company. It was anticipated that modifications of the redesign would take place during the implementation phase. The technical analysis included: a geographical layout and description of each operation; a breakdown of the wheel line into unit operations with a description of inputs, transformations, and outputs, and an analysis of key process variances and their interrelationships. The social analysis consisted of: a description of the work roles and tasks of the social system, the geographical relationship among work roles, an analysis of key variance control, and an examination of psychological needs for each work role. The analysis concluded by examining critical interfaces between the wheel line and the supply, user, and maintenance systems.

A redesign proposal based on this socio-technical analysis was formulated. The proposal was derived from the design criteria: a group of workers for a set of interrelated tasks. Two autonomous work groups were created for two sets of relatively interdependent tasks: one for the machining and drilling operations, and one for the polishing, degreasing, stamping, and packing operations. The concept of management was also reconsidered to include boundary management and open systems planning.

TRAINING WORKERS

While the machinery was being installed on the wheel line, workers were hired and a training program was started; its primary purpose was to introduce workers to the autonomous group method of production and to train them in the skills necessary to perform the wheel-line tasks. The initial trainees consisted of 12 workers divided equally between the two work teams. Each worker volunteered for the experiment and was chosen from a total of 76 volunteers on the criterion of seniority as specified in the sanction. Surprisingly, the seniority cut-off for the experiment was 32.4 years.

Two external consultants, an internal consultant, and the supervisor of the finishing operation conducted the training sessions. Daily, these sessions consisted of several hours of team development, followed by on-the-job skill training. Team development included team-building, communication, and self-management exercises; these activities contrasted sharply with the firm's traditional method of training which consisted primarily of skill training. Conversely, skill training was accomplished by the more common method of learning by performing. At about the middle of the training program, 12 additional workers were recruited and put through an abbreviated version of the training program. Also, a supervisor was added for the second-shift of operations.

The team development part of the training program was similar, in many ways, to an off-site team-building session: workers were encouraged to explore new ways of behaving and to discuss their feelings as well as the content of their interactions. Free from the constraints of a typical work day, individuals were given freedom to develop their work teams, to discuss the experiment, and to schedule rest breaks. This freedom, when combined with the consultants' discussions of the opportunities for self-management in the work groups, gave the workers a distinct impression that the experiment was not just a lot of talk or rumor, but rather a real chance to contribute to the management of their work lives. In fact, there were numerous instances where individuals felt so strongly about this "new way of working" that they discussed, for the first time, the content and conditions of their work at home. During this period, the question was frequently asked: "Why didn't the company start this approach 20 years ago?"

The skill training part of the program encountered two difficulties. First, it was anticipated that qualified machinists would bid on the machining team's jobs; however, to comply with the collective-bargaining agreement, membership in the experiment was based solely on seniority. This resulted in a group of machinist trainees with little, if any, experience in operating machines. This complicated, somewhat, the skill training as the supervisor had to spend more time than he had available to train workers; also, the equipment, which arrived late and was in poor condition, did not run well enough to allow for adequate training on all relevant machines. The second problem was the condition of the equipment. Based on the socio-technical analysis, it appeared to be in good condition. This was not the case. Much of the equipment was over 40 years old and arrived with no operating instructions, spare parts, or schematic diagrams for maintenance. When trainees inadvertantly abused the machines—as is often the case in skill training—the machines broke down. Since it frequently took hours and sometimes days to figure out how to repair the equipment, the skill training quickly degenerated into an unpredictable process. As the implementation phase of the experiment drew nearer, it became apparent that the machining team would not have the requisite skills to rotate members among all of the machines. Since contractual obligations with the automotive customers necessitated a rapid start-up of the wheel line, the training program could not be extended to allow for adequate skill training. Therefore, with a group of machinists not fully trained and a set of equipment in questionnable condition, the implementation stage of the experiment began.

IMPLEMENTING THE EXPERIMENT

At best, the start-up of a new operation is unpredictable. Problems of machine utilization, production-line balancing, quality assurance, and inventory control combine with the difficulties of developing new work roles and task structures to produce a series of tension-filled events. The finishing operation was no exception. The workers experienced technological and production problems that would boggle the mind of most industrial engineers. For instance, under the load of continuous performance, many machines broke down with the attendant problems of trying to repair them without the necessary knowledge or parts. Add to this such difficulties as poor wheel forgings, an inappropriate accounting system, inadequate inventory control, and a parade of guests observing the experiment, and the start-up process became more a test of workers' patience and stamina than of skill or training.

The production and technological problems had a major impact on the formation of the work teams. Rather than provide stability for developing the teams, the wheel line was so turbulent that the work groups were not able to develop stable role relationships or work patterns. As machines broke down unpredictably, members of the machinist team (Team A) were forced to move to other machines or tasks in a similarly unpredictable fashion. This problem was exaggerated by the fact that many machinists possessed limited skills, thus reducing their ability to move from one machine to another. Given the equipment problems, the quantity and quality of wheels that were machined for packaging by the second work group (Team B) varied drastically. This made it difficult to develop a stable rhythm to their work. Similar problems existed in regard to the condition of forged wheels that entered the wheel line. Because these wheels were a new product for the forgers of the company, they had initial problems meeting specifications. Since the quality of a forged wheel was often difficult to determine until it had gone through some of the machining operations on the wheel line, the feedback to the forgers was relatively slow. Thus, the input to the wheel line, in the form of forged wheels, was unpredictable and not easy to correct. Although we could add several pages about similar kinds of problems that impeded team development, our primary point has been made: the wheel line—its equipment and inputs, for example—was so disrupted that team formation and development was difficult.

As the weeks wore on, many of these problems were partially resolved. Given some stability, the work teams developed gradually toward responsible autonomy, but additional contracts for wheels added new problems. Thirty-two additional workers were assimilated into the wheel line to meet new production demands. Pressure for production shortened their training program; it also curtailed many of the experimental activities and additional skill training for the initial workers. Rather than slow the implementation process to work through these difficulties, management added more supervision to manage the larger scale of operations and to bring additional equipment and production processes on-line. This had the effect of reducing workers' opportunities for autonomy as the new supervisors frequently acted from a bureaucratic mentality. The consultants confronted this problem by attempting to resanction the experiment. This met with tacit approval from top management; yet under the strain for increased production, the resanctioning process was unsuccessful. In the face of these pressures, the implementation phase of the experiment ended, and the wheel line began the difficult task of meeting stringent production schedules with a relatively stable, but not fully autonomous, work structure.

EVALUATING THE EXPERIMENT

The experiment, like most new innovations, was neither a success nor a failure. If we examine the results of the experiment from the perspectives of each of the key participants—the workers, the managers, and the consultants—a better understanding of its affects is possible.

Workers' Perspective

The workers considered the experiment to be a step in the right direction, yet not a large enough step. They enjoyed the experience of being part of something new and of helping the company to survive in the wheel business. Here are some quantitative data that pretty much tell their story.

Research Design. Approximately one year after the start of the experiment, the experimental group and a control group were administered a questionnaire to ascertain the effects of the experiment. The comparison group consisted of those who had bid on experimental jobs, but who were rejected because of insufficient seniority. The logic behind this posttest research design is based on the assumption that the major difference between the comparison groups at the start of the experiment was seniority and any other variables that covary with seniority—e.g., age and education. Thus, if the questionnaire results are adjusted for initial differences in seniority, age, and education, significant posttest differences between the comparison groups are likely to be due to the experiment. Specifically, by using a covariance analysis and statistically adjusting for differences in seniority, age, and education, it is possible to attribute posttest differences between the two groups to the experiment.

Perceptions About Work Groups. Table 7 lists the results of the questionnaire regarding workers' perceptions about their work groups. Eight variables having to do with various facets of autonomous group functioning were measured on 7-point, Likert-type scales. The results of the covariance analysis—with seniority, age, and education as covariates—reveals that the experimental group is significantly higher on the question of "chance to see its task through from beginning to end" than the control group. On the other hand, the control group scores significantly higher than the experimental group on the autonomy dimensions related to "setting production goals" and "making quality decisions," as well as on the feedback variable of "knowing how well or poorly it is doing."

The data support the premise that the conditions for autonomous

Table 7

PERCEPTIONS ABOUT WORK GROUPS: COVARIANCE ANALYSIS

	Actual Means		Adjusted Means	
	Experi-mental Group	Control Group	Experi-mental Group	Control Group
1. My group operates within a well-defined work area it can call its own.	5.7	5.5	5.8	5.4
2. The members of my group have the necessary skills to do the overall job for which they are responsible.	5.0	5.7	5.1	5.5
3. My group has much control or final say over how it does its work.	4.5	4.5	4.2	4.8
4. My group has a chance to see its task through from beginning to end.	5.8	4.0	5.8*	3.9
5. My group has much to say in setting production goals.	3.6	4.2	3.3	4.4*
6. My group is responsible for the work it does.	6.0	6.2	6.0	6.2
7. My group makes most of the decisions about quality, such as when to pass or scrap the things it works on.	5.6	4.0	6.1	6.4*
8. My group knows much about how well or poorly it is doing.	4.0	5.5	3.8	5.7*

* < .05 level of significance.

Note: Experimental group n = 56, control group n = 40; negatively phased questions have been reversed for ease of interpretation.

group functioning were not implemented fully. Indeed, on the autonomy and feedback dimensions, the wheel-line workers mav well have been worse off than if they had stayed on their previous jobs. Although the results appear to be extremely negative, an alternative interpretation is that the wheel-line workers expected more autonomy and feedback than they actually received; these raised but unsatisfied expectations resulted in lower ratings relative to such expectations, although the wheel-line jobs may have provided more autonomy and feedback in some absolute sense.

Attitudinal Results. Table 8 presents the questionnaire variables having to do with workers' attitudes—need satisfaction, job involvement, and intrinsic motivation. Porter's Need Satisfaction Questionnaire (1961) was used to measure need satisfaction in five need areas corresponding to Maslow's (1954) need hierarchy: security, social, esteem, autonomy, and self-actualization. In each case, the score represents a need deficiency, the difference between "how much the need should be met on the job" and "how much it is presently met on the job." Seven-point scales were used to measure each question; the higher the score, the greater the need deficiency. Seven-point, Likert-type scales were used to measure items related to job involvement and intrinsic motivation. The former is concerned with how much the job situation is central to the person and his identity, while the latter is the extent to which higher-order need satisfaction is contingent upon performance. Job involvement was attained by six items presented by Lodahl and Kejner (1965) as good measures of this attitude; intrinsic motivation was obtained by four items shown by Lawler and Hall (1970) as distinct measures of this variable.

The data in Table 8 show that the experimental group has a significantly higher deficiency score in the security needs area than the control group. Otherwise, there are no significant differences between the comparison

Table 8

ATTITUDES: ANALYSIS OF COVARIANCE

	Actual Means		Adjusted Means	
	Experi-mental Group	Control Group	Experi-mental Group	Control Group
NEED SATISFACTION:				
Security Needs	1.9	2.4	2.7*	1.6
Social Needs	1.0	1.4	1.3	1.1
Esteem Needs	1.8	1.9	2.1	1.7
Autonomy Needs	1.8	2.4	2.0	2.1
Self-Actualization Needs	1.7	2.4	2.2	1.8
JOB INVOLVEMENT	4.5	3.7	4.3	3.9
INTRINSIC MOTIVATION	6.0	6.0	5.9	6.2

* < .05 level of significance.

Note: Experimental group n = 56, control group n = 40; the need satisfaction measures are deficiency scores, the higher the score, the greater the need deficiency.

groups on the remaining attitudinal variables. One interpretation of these results is that the experiment did not result in poorer attitudes; that is, the attitudes of the experimental workers were no worse than they were prior to the experiment. The higher deficiency score in the security needs area may be explained in terms of the insecurity experienced by those who entered the experiment. In other words, experimental workers did not have the security experienced in their previous work roles. A more pessimistic explanation of the data is that the experiment resulted in no anticipated increases in attitudes; in sum, workers would have been just as well off if they had been left alone.

Work-Related Behavior. Table 9 reports questionnaire and company data about work-related behavior—effort, performance, and withdrawal. Self-rated and supervisor-rated effort and performance were measured on seven-point scales, while the absence and tardiness data were obtained from company records for a six-month period. On all self-rated measures of effort and performance, the experimental group is significantly lower than the control group; the opposite is the case for the supervisor-rated measures. Absenteeism is significantly higher for the experimental group than the control group, while there is no significant difference in tardiness between the comparison groups. Although it is possible to attribute the wheel-line groups' lower self-rated effort and performance to the numerous equipment and production problems encountered during the implementation phase of the experiment, it is also plausible that these lower scores represent workers' assessment of their performance relative to what it should have been if the

Table 9

WORK-RELATED BEHAVIOR: ANALYSIS OF COVARIANCE

	Actual Means		Adjusted Means	
	Experi-mental Group	Control Group	Experi-mental Group	Control Group
Self-rated job effort	6.0	6.0	5.7	6.3*
Self-rated job performance	5.3	5.9	4.9	6.3*
Self-rated group effort	5.3	5.6	4.9	6.0*
Self-rated group performance	4.7	5.7	4.3	6.1*
Supervisor-rated job effort	5.3	5.0	5.2*	5.0
Supervisor-rated job performance	6.0	5.2	5.8*	5.4
Absenteeism	6.7%	4.6%	8.1%*	3.1%
Tardiness	2.7%	4.7%	3.1%	4.0%

* $<.05$ level of significance.
Note: Experimental group n = 56, control group n = 40.

groups were functioning properly. The higher absentee rate is likely to be due to the heavier physical and overtime demands placed on workers in the finishing operation in contrast to the control group workers. It is interesting to note that the supervisors of the experimental group rated their workers as putting forth greater effort and performance than the supervisors of the control groups. Since the supervisors of the finishing operation were under severe pressure to produce, it is likely that their ratings reflect, in part, their desire to appear productive to top management. The performance ratings, to be more specific, provided others with the impression that the supervisors were doing their jobs effectively. It is also likely that part of this higher rating is due to the workers' great diligence and effort during the implementation process.

Management's Views

The managers considered the experiment a qualified success. Though they agreed that the socio-technical approach had not been implemented fully, they gave the following reasons for their optimism:

1. The wheel line was implemented in record time for the X Company. It went from an empty building to a fully-producing operation in a matter of months.
2. For the first year of the experiment there were no lost-time accidents, an unbelievable feat given the high accident rate traditionally associated with the forging industry. In fact, the wheel line won an award for its accident record.
3. The wheel line was productive and profitable. When the equipment ran, productivity was above expectation. The wheel business went from 3 percent of the company's gross sales prior to automotive wheels to about 33 percent after the automobile wheel business. Also, the wheel line was run with fewer workers than were used by the subcontractor.
4. There were no union grievances on the wheel line, though the rest of the company was overloaded with such grievances.
5. Wheel-line workers frequently performed work, such as maintenance, outside of their union classification.
6. When necessary, workers from Group B chose to perform Group A's work; they refused the increased wages that this work provided until the wheel line was profitable.
7. When members of another union in the plant threw a wild-cat strike, the wheel-line workers were the only members from their union, which represented 95 percent of the work force, to cross the picket line. This was unheard of in the history of the company.
8. Members of the wheel line felt that their supervisor was being overworked during the implementation, phase of the experiment. When he lost over 60 pounds, starting from a slim build, they secretly asked the plant physician to put pressure on top management to lighten his work load. This was also without precedence in the company.

9. The parent company of the X Company, one of the largest firms in the world, put a picture of the wheel line on their annual report and included a write-up about this new approach within the report.

Consultants' Perceptions

The consultants saw the experiment as a considerable improvement over the preceding EST/DE study. Though the conditions for autonomous group functioning were less than ideal, the teams were able to function and to resolve many of their problems. Furthermore, the expansion of the wheel business helped the company to overcome its financial crisis. These favorable impressions were tempered, however, by several difficulties that diminished the experiment's positive effects. A retrospective examination of some of the more pertinent issues may add to our understanding of the socio-technical approach.

The Experiment as a Double Bind Situation. In many ways, the wheel-line experiment had the characteristics that Bateson and his colleagues ascribe to double bind situations (Bateson, 1972). Briefly, the double bind situation is derived from communication theory and is meant to describe a class of communicative disorders that may result in schizophrenia. Our intention here is not to imply that the experiment produced schizophrenic behavior, but rather to explore the possibility that the wheel line may have placed workers in a double bind situation, with the resultant anxieties and tensions.

From the perspectives of the individual caught in a double bind, such situations have the following characteristics:

1. A communicative relationship in which it is vitally important for the individual to discriminate as accurately as possible what sort of message is being communicated so that he may display appropriate behavior.
2. The individual is caught in a situation in which the other person in the relationship is expressing two contradictory orders of message.
3. And, the individual is unable to comment on the messages being expressed to correct his discrimination of what order of message to respond to.

Double bind situations occur frequently in normal relationships. Individuals caught in a double bind tend to respond defensively while experiencing a great deal of tension and anxiety. Bateson (1972) and his coworkers describe a typical double bind situation:

> One day an employee went home during office hours. A fellow employee called him at his home and said lightly, "Well, how did you get *there*?" The employee replied, "By automobile." He responded literally because he was faced with a message which asked him what he was doing at home when he should have been at the office, but which denied that this question was being asked by the way it was phrased. (Since the speaker

> felt it wasn't really his business, he spoke metaphorically.) The relationship was intense enough so that the victim was in doubt how the information would be used, and he therefore responded literally (p. 209).

The wheel-line experiment had many characteristics similar to a double bind situation. First, a communicative relationship existed between workers and management. From the perspective of workers, it was extremely important to understand management's message about the experiment so that appropriate behavior could be displayed in the experimental situation. Second, management expressed two contradictory orders of message. The primary message was that the experiment was sanctioned to provide workers with the protection necessary to experiment with the new work design. The secondary message, presented implicitly at a more abstract level, was that the finishing operation needed to produce as quickly as possible if the company was to survive in the new commercial market. Responses to these messages produced conflicting results: if workers chose to experiment with the autonomous group design, they were not able to produce as quickly as possible, but if they concentrated on production, they were not able to experiment fully with the team structure. Finally, workers were unable to confront these contradictory messages because they were of different orders of abstraction. The primary message was stated explicitly as the major reason for the experiment; indeed, the union and management sanctioned the experiment to provide protection for redesign. The secondary message, on the other hand, was at a more abstract level of communication—it was not stated specifically, but rather by non-verbal clues, such as gestures and tacit pressures to produce. Nonetheless, this more abstract message provided a larger context for denying implicitly the validity of the primary message. When received simultaneously, the two orders of message implied: "We expect you to experiment with the team design; yet, if you do, you will not produce at a rate that will allow us to survive."

Given this double bind situation, workers experienced much tension and anxiety in trying to respond appropriately. When they reacted literally to the primary message of experimentation, they encountered subtle forms of punishment concerning their productivity. For instance, attempts to develop team responsibility through group meetings received verbal support from management, followed by an injunction to schedule such meetings before or after regular working hours. Similarly, when workers became embroiled in production problems, management countered with pressures to proceed with the experimental process. Over time, the implicit clues of the secondary message for increased production were difficult to deny. Still, management persisted in their verbal support of the experiment, despite tacit demands for more wheels. As the illusion of experimental support wore thin, workers became increasingly hostile toward management's denial of the pressure to produce. Finally, additional wheel contracts placed such severe production demands on the wheel line that the experimental process was called to a halt. This terminated the double bind situation as the conflicting orders of message were reduced to only one level of communication: Produce! Although workers seemed relieved by this decision, their negative responses to the questionnaire may be an expression of this double bind situation.

Our discussion of the experiment in terms of the double bind situation raises an interesting question about management's conflicting messages: Why did they deny the reality that production was more important than socio-technical experimentation? The answer to this question becomes clear if we examine the assumptions underlying the experiment. The experiment was based on the expectation that the start-up process would proceed smoothly, thus allowing time and resources needed for experimentation. When equipment and production problems intruded on the experimental process, it was easier to proceed under this original assumption than to confront the harsh reality that experimentation was not feasible. Trist and his colleagues (1963) suggest that organizational members may cling to assumptions of ordinariness as a psychological defense against the realities of an unpleasant or nonordinary situation. In our case, management adhered actively to the assumption that the experiment would proceed normally despite evidence to the contrary. Given the amount of effort and affect tied to the experiment, it seems likely that this denial of reality was a psychological defense against the difficulties encountered, especially the growing pressures for production. Furthermore, management's unfortunate decision to purchase faulty equipment probably increased their anxiety about confronting the consequences of their mistake. Given this denial of reality, it is not surprising that management unintentionally sent two orders of conflicting message: the primary message, to experiment, based on assumptions of ordinariness and the secondary message, to produce, based on the need to meet stringent production schedules.

In retrospect, the wheel-line experiment placed workers in a double bind situation. Management's assumptions of ordinariness as a defense against the anxieties related to increased production pressures seem to have contributed to this situation. The conflicting orders of message, to experiment and to produce, resulted in hostility between workers and management. Perhaps a more realistic evaluation of the situation would have uncovered this problem and allowed it to be worked through. If so, this suggests that socio-technical experimentation may benefit from a "reality contract" whereby those involved in the experiment would spell out their assumptions and address possible inconsistencies. Reality contracts seem especially relevant in those experimental contexts susceptible to double bind situations: organizations facing nonordinary situations or not fully committed to a socio-technical strategy. Reality contracts would sensitize such organizations to the possibility of double bind situations by exploring the different orders of message likely to be communicated in such experimental situations. This would provide both workers and management with a realistic basis for sanctioning an experiment, and it would reduce the likelihood that conflicting orders of message confound the experimental process.

Maintenance Variance. An unexpected problem that plagued much of the experimental process was the faulty condition of the finishing equipment. In review, the analytical phase of the experiment failed to uncover these equipment problems. In fact, examination of the equipment at the subcontractor's plant led to the conclusion that it was in good order, with few maintenance variances. This raises the obvious question: How could the analysis be so faulty? Although we do not have an obvious answer to this question, there appears to be at least two possibilities. First, the subcontractor could have easily concealed the actual condition of the equipment. The maintenance analysis was done in a matter of hours, using visual observations and incomplete maintenance records. This explanation seems plausible in light of the subcontractor's desire to sell the equipment. Perhaps, we encountered a skillful used-equipment salesman. Second, the equipment may have been in poor condition, yet skillful workers kept it running. Our analysis did not take this possibility into account, nor did it anticipate the treatment the equipment would receive from the unskilled finishing workers. This is a pertinent point, since the equipment ran fairly well when it was treated properly. Although it is difficult to surmise, both of these explanations may account, in part, for the faulty analysis.

The failure to discover maintenance variances suggests that this part of the analytical process may have been neglected inadvertantly. The tendency may be to assume that the mechanical parts of the technological system are rudimentary in contrast to the layout of the production process and the other technological variances. If so, we would expect the detection of maintenance variances to be a problem, especially in those instances where the equipment is complex, or where a preventative maintenance program is not followed strictly, or where the equipment is new to the organization. Given this potential problem, it seems prudent to call attention to the maintenance analysis and to underscore the possibility that it may be disregarded unintentionally.

Technology and Work Culture. The wheel line represented a new technology for the X Company. The finishing tasks comprised a long-linked or mass production technology, since standard operations were performed repetitively on a similar product. This contrasted sharply with the firm's primary technology which was a job shop or unit operation, consisting of short production runs on many different products. The decision to embed a mass production technology within a job-shop organization produced an unexpected problem related to the work culture of the company. The prevailing work culture of the firm was geared to the requirements of a forging technology. Here, the human element dominated the production process as teams of skilled forgers determined both the pace and quality of production. Because forging is as much an art as a science, much time was spent adjusting machines and setting up each production run. In many ways, this work culture was man-centered, since human utilization was relatively more important than machine utilization.

Given this predominate work culture, both workers and management had a difficult time adjusting to the demands of the wheel line. Specifically, the finishing tasks were machine-paced, hence they required a culture that supported machine utilization. This also placed a new value on time as productivity was dependent upon the continuous operation of machinery.

The tendency to view the wheel line from the perspective of a man-centered work culture led to several unproductive decisions. A prime illustration concerned the staffing requirements of the machinist team. Rather than provide back-up workers to replace absentees, management chose to operate with a minimal crew. This resulted in reduced productivity as machines often sat idle while needed workers were absent or tardy.

The problems inherent in applying an existing work culture to a work system that demands a different set of assumptions are not overcome easily. The premises that underlie work cultures are often so ingrained in organizational members that they are difficult to detect and still harder to change. Currently, the analytical models do not place enough emphasis on examining work cultures as they relate to experimental units. This suggests that socio-technical experimenters may have to devise their own method for assessing the work culture of the organization, especially in those instances where the experiment constitutes a significant departure from existing organizational technologies and practices. Conceivably, the initial scanning could incorporate an examination of the organization's work culture. Relevant information might include: (1) the prevailing customs and routines influencing behavior, (2) the current accounting procedures setting the parameters for regulation and control, (3) the organizational structures and technological layouts determining the rationality of productive effort, and (4) the managerial philosophy setting the tenor for worker/management relationships. This information, when compared to the needs of the experimental system, may alert experimenters to potential problems in implementing a redesign. Furthermore, examination of the work culture may uncover those organizational factors that need to be changed before socio-technical experimentation is considered.

Training Program. The wheel-line experiment, like many socio-technical redesigns, required much skill and team learning. For workers, the training program provided experience to develop work teams and to perform the requisite tasks. Unfortunately, the attention paid to workers distracted from the training needs of management. This proved to be a costly error as management did not have sufficient knowledge or skills to implement and to manage the work teams. During the early stages of the experiment, this lack of training was not evident; here, supervisors joined actively in theory sessions and in impromptu discussions about the socio-technical approach. Although many questions were raised, these discussions gave the general impression that management understood and supported this perspective. Once the implementation process started, supervisors experienced great difficulty in applying socio-technical theory while trying to resolve production problems. This situation was made worse by increased demands for production and by informal ostracism from other supervisors who resented this "country club" approach. As pressures mounted, managers began to question the autonomous group method and to behave in a more traditional style. When confronted by consultants, they countered with such statements as: "This stuff sounds good, but it doesn't work a damn on the shop floor!" Belated attempts to furnish needed training met with tacit resistance as these were seen as distracting from the work on the line.

The failure to respond to managerial training needs suggests that this may be a neglected aspect of socio-technical experimentation. Indeed, our experience shows that supervisors are frequently left to their own wits to determine how to behave in experimental settings. Although this intuitive approach may suffice in some instances, a more formal training program appears needed in those situations requiring new forms of behavior.

This is an important point, for autonomous group redesigns require a style of leadership that is quite different from traditional forms. Here, the supervisor must provide workers with sufficient direction and support to develop toward responsible autonomy, and he must manage the work system's boundary. These tasks demand a thorough knowledge of the work system and its environment in addition to a familiarity with social science practice, especially interpersonal and group dynamics. Learning to perform these tasks requires a combination of theory and practice, and preferably, such training would be in a protected setting where individuals can freely experiment with new behavior. Similarly, on-the-job training may be more effective if careful attention is paid to the learning opportunities available in the situation. Debriefing sessions and scheduled meetings with knowledgeable resource people may serve this function. Finally, it is important to note that managers often mistake the autonomous group approach for a laissez-faire style of management. What this "hands off" interpretation fails to recognize is that autonomy is possible only within clearly defined boundaries of discretion, and it is the responsibility of management to provide such boundaries.

In summary a restrospective examination of the wheel-line experiment raised some pertinent issues that may affect the outcome of socio-technical experimentation. First, the experiment appeared to place workers in a double bind situation. Two orders of conflicting message, to experiment and to produce, led to much anxiety as workers were unable to respond in an appropriate manner. A major cause of this untenable position was management's assumptions of ordinariness as a psychological defense against the reality of certain production problems. A possible resolution to such double bind situations is a reality contract whereby those involved in the experiment would spell out their assumptions and address possible inconsistencies. This would reduce the probability of different orders of conflicting message and it would provide experimenters with a realistic basis for sanction. Second, the socio-technical analysis failed to detect important maintenance variances, pointing out that this part of the analysis may be overlooked inadvertantly. Third, the introduction of a mass production technology into a job shop organization produced an unexpected problem related to the work culture of the company. The assumptions behind the prevailing man-centered culture were inappropriate to the operation of the wheel line which demanded a machine-centered culture. Since socio-technical analysis may neglect the impact of an organization's work culture on the experimental system, this should probably be included in the initial scanning of the organization. This would alert experimenters to potential problems in implementing a redesign, and it would reveal those organizational factors that may have to be changed before experimentation is considered. Finally, the wheel-line experiment failed to respond appropriately to managerial training needs. Based on the style of leadership needed to support and to direct autonomous work groups, increased attention should be given to managerial training, both on-the-job and in off-site settings.

CONCLUSION

The wheel-line experiment represents an attempt to apply socio-technical theory on the shop floor. The specific circumstances surrounding the experiment suggest that this is a stringent test of this strategy for organizational change. In the face of severe economic pressures, any form of experimentation is likely to encounter unexpected difficulties. This raises the question as to whether this approach is responsive to such situations. Though there is no clear answer, our experience suggests that socio-technical theory is relevant only if applied experimentally. Put differently, the application of this approach requires a responsiveness to the contingencies arising in the work setting. Efforts to experiment in a mechanical manner present undue risks as emergent problems are likely to render standard procedures obsolete. The wheel-line experiment underscores the need to adjust continually to the situation. The major problems encountered during the study show clearly that sound theory is not enough; one must also understand the organization and tailor the experiment to its needs. This demands informed judgment and an appreciation for the process of experimentation: a method guided by theory and practice in which the results can never be known in advance.

10

THE WORK OF MANAGEMENT

The management of work constitutes a strategy for diagnosing, designing, and developing socio-technical systems. These three D's, as it were, represent much of our knowledge about managing man's relationship to technology. Diagnosing involves an understanding of the work system: its internal dynamics, both social and technical, and its external relationships with the environment. This knowledge guides work system design. Here, we are concerned with jointly optimizing the social and technical components and creating an effective match between the system and its environment. Finally, development includes providing for a succession of suitable environments enabling the system to grow. When all three processes—diagnosing, designing, and developing—are managed integratively, work is experienced as humanly satisfying and productive.

Whereas much of the management of work has focused on the work structure that relates people to technology, little has been said about the *work of management*, a function that arises with the extension of the socio-technical concept to the whole organization. At this level, the organization is an internally differentiated entity in which the work structure or production subpart relates to other parts, such as procurement, marketing, personnel, and maintenance. The integration of these components into an organized whole requires a new order of management, one concerned with the organization as an extended social system. Figure 1 illustrates the emergence of this management function in relation to three levels of organizational responsibility—the technical, the managerial, and the institutional (Parsons, 1960). At the technical level, people relate to technology to perform the organization's primary task. The management of work as outlined in this book is most appropriate at this level, since it is concerned with this social and technical relationship. Indeed, much of what is commonly referred to as work in organizations occurs at this level. As we move upward toward the managerial and institutional levels, work loses its distinctive technological character and becomes more of a social process; here, people's relations to each other integrate the separate organizational parts and relate them to the wider environment. The dynamic interactions existing among the three organizational levels demand a management function responsive to this differentiated whole. This is the work of management, a necessary complement to the management of work in organizations.

Like other forms of work, the work of management is based on an agreement between two or more persons to perform a stated task. In this instance, the work agreement applies to individuals' relationships to the total organization; that is, it relates members to an organized whole with a

Figure 1

THE MANAGEMENT OF WORK AND THE WORK OF MANAGEMENT

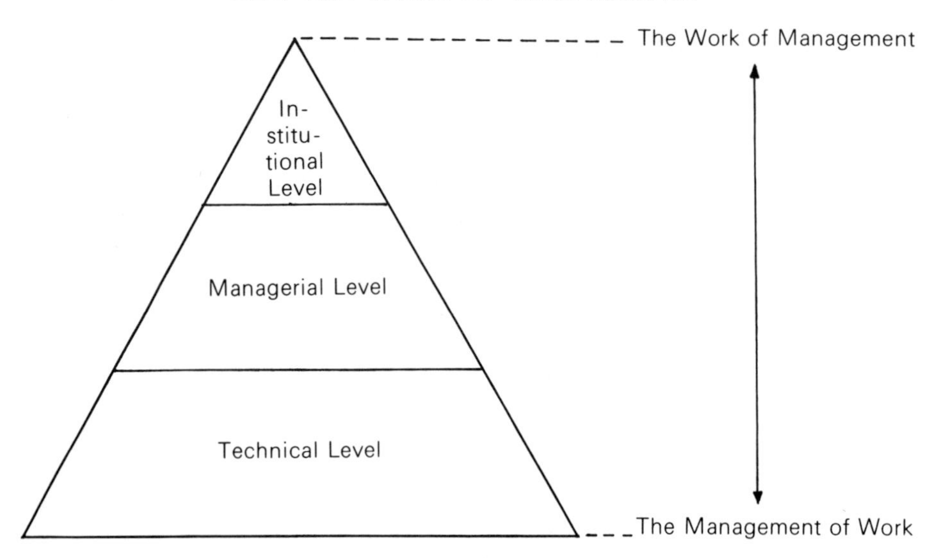

specific purpose. Organizationally, an interpersonal definition of work results in four kinds of work: prescribed, contractual, discretionary, and emergent. Each form of work may be considered a specific management function, expressive of the work of management. Since traditional socio-technical concepts have not paid adequate attention to the work of managers, our purpose is to use the concepts of work described in this book to explore what the work of management comprises from this perspective and provide a preliminary extension of the socio-technical approach to current management practice.

PRESCRIBED WORK: THE MANAGEMENT OF AUTHORITY

The management of the organization as a social system presents a major problem in constraining members' behavior to those actions necessary for goal achievement. Given the inherent irrationality of the human component, management must set limits to the use of personal discretion. This provides individuals with clear boundaries for judgment, and it furnishes the organization with the rationality needed to operate as a differentiated entity. Prescribed work represents this function of management, since it involves the determination of work by others. Here, we are concerned with structuring the roles of the organization so that individuals have clear boundaries defining the use of discretion. The need for orderly and rational task accomplishment and review makes the role structure in an organization hierarchical; superior role holders prescribe the work of subordinates. The ability to determine others' work is based on the authority ascribed to organizational roles. The legitimate use of this authority rests with the "authority of the superior," an institutionalized set of expectations concerning people's willingness to comply to superiors' prescriptions. Since authority sets limits to the use of discretion, the *management of authority* is necessary to bound the use of power and the discharge of responsibility in the firm and to structure rationally the roles of the organization.

Bounding Power and Responsibility

The management of authority is concerned with delegated power (Emery, 1959). The hierarchical structure of the organization requires that each level specify the power and hence discretion of the next lower level. This enables the firm to control its extended parts and to direct its members toward a common goal. Since delegated power is distributed throughout the organization, restrictions on its use are necessary if it is to serve the purposes of the organization. To prescribe bounds to the use of discretion, each managerial level must clearly specify over whom or what activities power is to be exercised as well as how it is to be used. Furthermore, since power without responsibility leaves the organization open to the arbitrary use of that power, power and responsibility must be congruent. Thus, managers must also explicate to whom the subordinate is responsible, what he is responsible for, and over what time period and how his discharge of responsibility will be judged. When power and responsibility are bounded clearly within organizational roles, the management of authority provides the organization with specified bounds to the use of discretion.

Structuring Organizational Roles

In addition to defining clearly both power and responsibility, the management of authority must determine the breadth or scope of members' discretion. This may be understood in terms of the optimal structuring of the organization's roles (Emery, 1959). In defining lines of power and areas of responsibility, managers structure the roles of the firm. Since our concern is to structure roles to account for their open system properties, managers must provide for optimal degrees of discretion if individuals are to exercise their self-regulating capacities. Over structuring roles, by narrowly defining lines of power and areas of responsibility, is likely to interfere with this self-regulating property. This reduces the behavioral variability needed to adjust to changing circumstances. On the other hand, under structuring roles is likely to leave members with an insufficient boundary for the exercise of discretion and lead to confusion and to a lack of system integration. Given the need to establish an optimum arrangement of organizational roles, managers must use considerable judgment in determining the breadth of members' discretion; such judgment, if it is to enhance the organization's capacity for self-regulation, must be responsive to the open system properties of role holders.

CONTRACTUAL WORK: THE MANAGEMENT OF COMMITMENT

Much of this book concerns individuals' commitment to work. The design of work to meet people's needs is a powerful source of task commitment. At the organization level, however, commitment shifts from the task to the organization as a social system. Here, commitment constitutes an individual's willingness to interact with others toward attaining the organization's goals. Since this form of commitment derives from people's interactions, it involves the exchange of commitment among organizational members. Contractual work depicts this process; it includes the mutual determination of work through the exchange of commitment. Contractual work rests on the "authority of the contract," an agreement to interact in a stated way. This agreement commits individuals to each other through mutually determined behavior, the basis for organizational commitment. The *management of commitment* is concerned with two related processes: (1) agreements among organizational members and (2) the distribution of equity or justice in the social system.

Agreements Among Organizational Members

The management of commitment involves agreements among organizational members to interact in stated ways. Whereas this results in organizational commitment, it is not commitment to some abstract entity, but rather to others toward a common goal. This is a pertinent point, for organizational commitment is a social process binding people to each other through mutual agreements. Traditionally, organizations base their agreements on such tangible factors as wages, hours, and fringe benefits. Members agree to behave in a certain manner in exchange for these rewards. While this results in some minimal level of commitment, it is not sufficient to bind people fully to the organization.

Other grounds of agreement such as trust, loyalty, and comradeship, are powerful sources of commitment to a social system. Unfortunately, these more intangible bases of commitment are frequently left to the informal dynamics of the organization. The classical Hawthorne studies provide a lucid example of how workers can commit themselves to informal group norms. If such commitment is to serve the purposes of the organization, it must be institutionalized in the formal agreements among organizational members. This requires an explicit concern for the full range of agreements that tie individuals to each other; a willingness to give and to receive commitment on all of its dimensions.

Managers can institutionalize both formal and informal bases of agreement by being more cognizant of how and to what extent members exchange commitment. This would provide a formal means for improving the process through which members agree to certain behavior and hence commit to each other. Furthermore, since the exchange of commitment tends to be mutually reinforcing, managers must realize that commitment is gained by giving it. This exchange, when formalized in the interactions between superiors and subordinates, would likely result in a more broadly based commitment to the firm.

The Distribution of Equity

The need to institutionalize commitment in the formal agreements among organizational members can be understood in terms of equity or justice. Here, equity refers to individuals' feelings of fair treatment relative to others in the organization. People are likely to exchange commitment if they experience such exchange as equitable. Equitable exchange provides the organization with a sense of justice in the formal agreements that bind members to each other. This may be seen in reference to wage or salary agreements. Individuals are likely to commit their services to the organization if they perceive that their reward, pay in this instance, is equitable—that it is commensurate with their contribution and equal to the pay of others who perform similar work.

When the economic rewards of the firm are distributed equitably—that is, when rewards correspond to the allocation of authority and responsibility—the organization assures justice in its reward structure. To sustain justice, the enterprise must have an appellate process whereby members can appeal to a higher authority if they perceive that they are being treated unfairly. The formal grievance procedure that exists in most organizations represents this appellate function, at least for those inequities covered by the formal contract of employment. The appellate process ensures equity by enforcing mutual compliance to the work agreement. Without such enforcement, the organization is open to individuals' capricious interpretation of what is right and wrong.

While organizations take elaborate steps to provide justice in their formal agreements, they do little to provide equity in their more informal bases of agreement, such as trust and loyalty. Since these are significant sources of commitment, it is important to assure equity in this area. To provide for fair treatment on these dimensions, managers must expand the appellate process to include those more implicit grounds for agreement. This would likely involve a redefinition of justice throughout the organization; an extension of equity to cover both formal and informal exchanges of commitment. An explicit concern for just treatment in the interactions that bind people to the organization would provide individuals with a formal means to secure equity in their relationships. This would be a major step toward institutionalizing commitment to the firm.

DISCRETIONARY WORK: THE MANAGEMENT OF DECISION-MAKING

This form of work involves the use of discretion within limits set by prescribed work. Discretionary work goes beyond mere physical work to include the judgment and choice inherent in decision-making. Consideration of the organization as a social system raises the issue of multiple sources of decision-making. This demands a decision-making function enabling the organization to manage the use of discretion over an extended role structure. Decisions made in one part of the system must be consistent with those made in other parts if the firm is to operate as a unified whole. Furthermore, the consequences of different courses of action must be congruent if decisions are to contribute to a common goal. Since these issues transcend single components, the decision-making process must be managed from the level of the total organization.

Discretionary work represents the decision-making capacity of the organization; it receives its legitimization from the "authority of the sanction," a formalized set of prescriptions to the use of judgment. These prescriptions provide clear areas for discretion and discharge of responsibility. The *management of decision-making* involves the exercise of judgment within these boundaries. Specifically, this function is concerned with discretion over four elements in the social system: (1) information, (2) political processes, (3) succession, and (4) norms and values.

Information

Information is the sin qua non of organization; it integrates the parts of the system and provides individuals with knowledge needed for decision-making. Since information serves these functions through the communicative processes of the social system, management must provide for effective communication throughout the extended role structure. Specifically, to provide role holders with sufficient knowledge to exercise power and to discharge responsibility, information must be coordinated to the allocation of power and responsibility in the role structure (Emery, 1959). Furthermore, because unrestricted communication burdens the selective responses of role holders, restriction of sending and receiving information is necessary. To ensure that information is responsive to the needs of decision-makers, managers must specify clearly what information is needed, how it is to be obtained, to whom it is to be sent, and how it is to be used. This provides clear lines of information and enables decision-makers to apply relevant knowledge to their problems.

Political Processes

Social systems are inherently subject to the personal exercise of discretion beyond that specified in the formal role structure of the organization. These political processes involve personal agreements that bind people informally to certain courses of action. This wheeling and dealing, as it were, is a personalized response to the use of discretion in organizations. Because it involves implicit agreements that are not formalized publically, individuals are personally rather than organizationally responsible for this use of discretion. Political processes are a particularly inviting way to discharge prescribed work. Individuals can circumvent many of the formalized procedures for decision-making by gaining informal support for a decision before going through the specified channels of sanction. The well-used strategy of collecting votes or calling in one's markers as a prelude to the formal decision-making process is an example of the expediency of the political process.

The personalized character of the political process makes it especially susceptible to abuse. Members can personally gain power well beyond that ascribed to their organizational roles. Since this power is without formal responsibility, organizations cannot control its use easily. Thus, dependence on the political process as a means to decision-making places the organization in a precarious position. While managers cannot formalize all uses of discretion, they may curb its more flagrant abuses by making explicit the premises underlying decision-making. This provides members with an ethical basis for the exercise of judgment. Furthermore, an ethical code for decision-making is a powerful source of morality to the exercise of personal discretion in the organization.

Succession

Given the organization's hierarchical role structure, succession or advancement through the hierarchy is needed to ensure that roles are replenished and to provide individuals with paths to development. The exercise of discretion in this area is especially troublesome, since superiors must make personal judgments about who advances and who does not in the organization. Furthermore, since we are dealing with the total organization, these decisions must be integrated through time so that there are no fluctuations in the flow of personnel through the firm. To overcome these problems, organizations employ intricate procedures for rationalizing the succession process. Formal mechanisms, such as personnel audits and promotability reviews, provide some objectivity to what is otherwise a highly personalized decision-making method. Indeed, the familiar adage of "who you know" may not be far wrong. The quest to objectify these decisions invariably reduces succession to an evaluative process, where present role behavior is the criterion for advancement. This leads members to focus on the formal requirements of the role rather than on their development in the role. Thus, succession becomes more concerned with assessing current role bahavior than developing future role behavior.

Notwithstanding the need to rationalize methods for succession, managers may improve considerably the developmental opportunities associated with career advancement. This requires a dynamic view of people and the roles they occupy. Rather than conceive of organizational roles as static positions, they may be seen more dynamically as points on the career paths of individuals. This adds an important dimension to the evaluative element of succession; people are not only assessed in respect to role behavior, but the role itself is evaluated in regard to its opportunities for developing individuals' career paths. Thus, managers can greatly improve the developmental aspects of succession by explicating clearly people's location in their career paths, what learning is needed for advancement, and how the present role can provide such learning.

Norms and Values

Much of the work of management is concerned with formal procedures or mechanisms for integrating members into a unified social system. A powerful source of integration is the norms and values that individuals hold by virtue of membership in the organization. Here, we are referring to publically sanctioned rules of conduct and modes of evaluation that guide behavior and determine preference for specific end-states. These forms of information constitute the culture of the organization; that is, they determine how people structure reality and interpret what is consistent with the organization's mode of operation. If the organization is to function as an extended social system, members must share a consistent set of norms and values. These help to define the organization and to establish working procedures for the coordination and discipline of members' contributions (Olmsted, 1967).

While the power of informal norms is well documented, our concern here is the officially sanctioned rules of conduct and evaluative preferences of the organization. A major function of management is to explicate the organization's norms and values and to maintain them in the social system. This is necessary to ensure a sufficiently shared view of reality and to interpret the behavior and outcomes needed for goal achievement. Since norms and values are authoritative guides to the use of discretion, organizations cannot leave their determination to the informal dynamics of the social system. This would subject the organization to variable interpretations of how to discharge discretionary work. Indeed, many of the conflicts that permeate existing organizations are the result of enclaves of competing norms and values. To overcome this segregative tendency, managers must specify clearly the organization's norms of behavior and values for determining what are the socially preferred goals and conduct. If these prescriptions are to be maintained without undue coercion, they must be consistent with the prevailing norms and values of the wider culture. This provides the organization with a legitimate basis of conduct, a match between individuals' internalized norms and values and those of the firm.

EMERGENT WORK: THE MANAGEMENT OF ENVIRONMENT

Emergent work originates from the multitude of environmental forces operating in the work setting. Since this form of work is not determined directly by those involved in the work agreement, it rests on the authority of the environment, an external set of forces that compels response through its impact on the organization. In contrast to the other forms of work—prescribed, contractual, and discretionary—emergent work is external to the organization. It involves the relationship of the organization to the wider environment and requires a management function that is responsive to external conditions. The *management of environment* represents this function and it involves the determination of policy and the ecology of the organization.

Policy

The management of environment represents the organizations' attempt to control those environmental exchanges necessary for survival and growth. Given this interdependence, the internal dynamics of the firm must somehow reflect these external realities. Emery (1959) views this relationship in terms of the development of policy—"the central feature of the policy of the enterprise being a determination of what ends it shall pursue with the means actually or potentially available" (p. 42). The determination of ends delineates those points at which the organization is dependent upon its environment; thus, it must reflect, in large part, the demands of the environment. To manage this problem, policy must specify the mission or objectives of the firm in light of these external conditions. This requires a clear delineation of the organization's primary task—the conversion process that the organization must perform if it is to survive. At the organizational level, the primary task may be considered the strategic objective of the firm. Here, "the strategic objective should be to place the enterprise in a position in its environment where it has some assured condition for growth" (Emery and Trist, 1960, p. 97). This provides the boundary conditions for continued forms of environmental exchange. Furthermore, specification of the primary task or strategic objective furnishes the organization with stability and direction in the face of environmental uncertainty. Without clear policy, the organization is subject to a rudderless course as it negotiates its environment.

Ecology of the Organization

Whereas policy defines the organization's primary task, the environment itself must be managed if the firm is to consummate its objective. This requires knowledge of the dynamic forces in the environment as well as a capacity to influence these forces in favorable ways. Understanding the environment starts with the realization that it possesses its own set of systemic properties apart from the organization. Emery and Trist (1965) describe those processes through which parts of the environment become interconnected as the "causal texture" of the environment. Given this causal texture, the organization cannot simply concern itself with direct forms of exchange with its environment. Rather, it must appreciate that changes in one part of the environment may amplify greatly to other parts in ways not always predictable. Thus, the organization's actions may trigger changes in the environment leading to numerous unintended consequences; conversely, autonomous changes in the environment may impact the firm in similarly unintended ways.

An appreciation of the environment as a richly connected field reflects an ecological stance toward management. This means that the firm must consider itself as only one part of a vast complex of other organized entities that together constitute society. Since the actions of one member of society may adversely affect the other parts, the organization must sooner or later come to terms with its responsibility to this wider system. An ecological perspective accounts for this responsibility through managing the organization as a viable part of society. Managers must assure that the organization's tasks are responsive to society's needs, that its internal activities embody the norms and values of the wider culture, and that it promotes and enhances positive values both inside and outside its boundaries. It is only when managers take this positive stance that it can be said that they manage their environment responsibly.

CONCLUSION

This brief discussion of the work of management underlies the initial premise that the organization constitutes a social system of interdependent parts. Although the presentation is meant to be suggestive about the function of management at this level, it provides a preliminary extension of the socio-technical concept to the whole organization. Additional conceptualization and research is needed to develop this important aspect of management practice. Indeed, a logical extension of socio-technical theory and research to the total organization would be a valuable complement to the work outlined in this book.

The work of management represents a philosophy of organization; it is a set of assumptions about organizations as open systems, continually faced with uncertainty yet demanding rationality. Given this perspective, the organization must manage an inherent dilemma: how to bring structure or certainty to the whole, while remaining open to change and uncertainty. Like all dilemmas, there is no easy resolution to these contradictory demands. Attempts to bring structure or integrality to the organization invariably thwart its capacity to cope with external complexity and change. Traditional management practice has tended to settle the problem of adaptability at too low a level; it has opted for certainty at the expense of flexibility. The philosophy of organization presented in this book is an attempt to rectify this imbalance. The four kinds of work—prescribed, contractual, discretionary, and emergent—constitute phases on the continuum from certainty to uncertainty. Movement through these phases, from prescribed work to emergent work, represents the continual opening up of the organization to the uncertainty of the wider environment; conversely, movement in the opposite direction, from emergent work to prescribed work, is the closing down of the organization to the certainty needed for task performance. The management of the four forms of work enables the organization to provide certainty for its primary task, while managing its openness to the wider environment.

The management of work and the work of management are complementary functions for the management of organizations. Whereas the former is concerned with diagnosing, designing, and developing organizational parts, the latter integrates these parts into a unified whole and relates this whole to the wider environment. Both functions derive from socio-technical theory and practice; both view the organization as an open, socio-technical system. The similarities between the two make it difficult to differentiate between them. Indeed, the major difference is one of perspective: The work of management focuses on the total organization, while the management of work concentrates on the parts within it. In practice, this differentiation tends to become obscure as individuals develop their capacity to manage wider aspects of the total organization. Hopefully, this book contributes to the growth of this capacity and towards providing individuals with a deeper engagement with the organization by developing people's full potential for human enrichment and performance.

BIBLIOGRAPHY

Ackoff, R. L. "The Evolution of Management Systems." *Canadian Operational Research Society Journal*, 1970, *8*, 1–13.

Angyal, Andras. *Foundations for a Science of Personality.* New York: The Viking Press, 1941.

Argyris, C. *The Applicability of Organizational Sociology.* London: Cambridge University Press, 1972.

Aronoff, Joel. *Psychological Needs and Cultural Systems.* Princeton: D. Van Nostrand Company, Inc., 1967.

Ashby, W. Ross. *An Introduction to Cybernetics.* New York: John Wiley and Sons, 1966.

Atkinson, J. W. *An Introduction to Motivation.* Princeton: Van Nostrand Reinhold, 1964.

Bateson, Gregory. *Steps to an Ecology of Mind.* New York: Ballantine Books, 1972.

Bennis, Warren G. *Organization Development: Its Nature, Origins, and Prospects.* Reading: Addison-Wesley Publishing Company, 1969.

Blauner, R. *Alienation and Freedom: The Factory Worker and His Industry.* Chicago: University of Chicago Press, 1965.

Blood, Milton R., and Charles L. Hulin. "Alienation, Environmental Characteristics, and Worker Responses." *Journal of Applied Psychology*, 1967, *51*(3), 284–290.

Brown, Wilfred. "What Is Work?" *Glacier Project Papers.* Ed. Wilfred Brown and Elliot Jacques. London: Heinemann, 1965, pp. 54–73.

Buckley, Walter. *Sociology and Modern Systems Theory.* Englewood Cliffs: Prentice-Hall, 1967.

Campbell, Donald T., and Julian C. Stanley. *Experimental and Quasi-Experimental Designs for Research.* Chicago: Rand McNally and Company, 1966.

Clark, James V., and Charles G. Krone. "Towards an Overall View of Organization Development in the Early Seventies." *Management of Change and Conflict.* Ed. John Thomas and Warren Bennis. Baltimore: Penguin Books, 1972.

Conant, Eaton H., and Maurice D. Kilbridge. "An Interdisciplinary Analysis of Job Enlargment: Technology, Costs, and Behavioral Implications." *Industrial and Labor Relations Review*, 1965, *18*, 377–395.

Cummings, Thomas G. "Socio-Technical Systems: An Intervention Strategy." *Current Issues and Strategies in Organization Development.* Ed. Warner Burke. New York: Behavioral Publications (in press).

Cummings, Thomas G., Roy Glen, and Edmond Molloy. "A Methodological Critique of 57 Selected Work Experiments." Case Western Reserve University. Submitted for publication, 1976.

Davis, Louis E. "The Coming Crisis for Production Management: Technology and Organization." *International Journal of Production Research*, 1971, *9(1)*, 65–82.

————. "The Design of Jobs." *Industrial Relations*, October, 1966, *6(1)*, 21–45.

————. "Job Design and Productivity." *Personnel*, *33*, 1957a.

————. "Toward a Theory of Job Design." *Journal of Industrial Engineering*, 1957b, *8*, 305–309.

Davis, Louis E., and Eric L. Trist. "Improving the Quality of Work Life, Experience of the Socio-Technical Approach." A background paper commissioned by the U.S. Department of Health, Education and Welfare for a report: *Work in America*, June 1972.

Emery, F. E. "Characteristics of Socio-Technical Systems." Tavistock Institute of Human Relations, Doc. No. 527, 1959.

————. "The Democratization of the Workplace." *Manpower and Applied Psychology*, 1969, *1*(*2*), 118–129.

————. "The Next Thirty Years: Concepts, Methods and Aspirations." *Towards a Social Ecology*. Ed. F. E. Emery and Eric L. Trist. London: Plenum Press, 1973.

————. "Some Hypotheses About the Way in Which Tasks May Be More Effectively Put Together to Make Jobs." Tavistock Institute of Human Relations, Doc. No. T. 176, May 1963.

Emery, F. E., and Einar Thorsurd. *Form and Content in Industrial Democracy*. London: Tavistock Publications, 1969.

Emery, F. E., and E. L. Trist. "Socio-Technical Systems." *Management Science, Model's and Techniques*, Vol. 2 Ed. C. W. Churchman and M. Verhulst. Elmsford, New York Pergamon, 1960, pp. 83–97.

Emery, F. E., and E. L. Trist. "The Causal Texture of Organizational Environments." *Human Relations*, 1965, *18*, 21–32.

Feibleman, J., and J. W. Friend. "The Structure and Function of Organization." *Philosophical Review*, 1945, *54*, 19–44.

Festinger, Leon. *A Theory of Cognitive Dissonance*. Evanston: Row, Peterson, 1957 (reprinted by Stanford University Press, Stanford, California, 1962).

Hall, A. D., and R. E. Fagen. "Definition of System." *General Systems, I*, 1956, pp. 18–28.

Herbst, P. G. "Situation Dynamics and the Theory of Behavior Systems." *Behavioral Science*, 1956.

————. "Socio-Technical Unit Design." Tavistock Institute of Human Relations, Doc. No. T. 899, 1966.

Herzberg, F. *Work and the Nature of Man*. Cleveland: World Publishing, 1966.

Jaques, Elliott. *The Changing Culture of the Factory*. London: Tavistock Publications, 1951.

————. "National Income Policy." *Glacier Project Papers*. Edited by Wilfred Brown and Elliott Jaques. London: Heinemann, 1965, pp. 237–245.

Lake, Dale, Mathew B. Miles, and Ralph B. Earle. Jr. (eds.). *Measuring Human Behavior*. New York: Teachers College Press, 1973.

Lawler, E. E. "Job Design and Employee Motivation." *Personnel Psychology*, 1969, *22*, 426–435.

Lawler, Edward E., and Douglas T. Hall. "Relationship of Job Characteristics to Job Involvement, Satisfaction, and Intrinsic Motivation." *Journal of Applied Psychology*, 1970, *54*, 305–312.

Lodahl, T. M., and M. Kejner. "The Definition and Measurement of Job Involvement." *Journal of Applied Psychology*, 1965, *49*, 24–33.

Lorsch, J., and J. Morse. *Organizations and Their Members: A Contingency Approach*. New York: Harper and Row, 1974.

March, James G., and Herbert A. Simon. *Organizations*. New York: John Wiley and Sons, Inc., 1958.

Martin, Michael. "Operational Research; A New Discipline." *Management Thinkers*. Ed. A. Tillett, T. Kempner, and G. Wilis. Baltimore: Penguin Books, Inc., 1970, pp. 140–165.

Maslow, A. H. *Motivation and Personality*. New York: Harper Bros., 1954.

Miller, E. J. "Technology, Territory, and Time: The Internal Differentiation of Complex Production Systems." *Human Relations*, 1959, *12*, 243–272.

Miller, E. J., and A. K. Rice. *Systems of Organization, the Control of Task and Sentient Boundaries*. London: Tavistock Publications, 1967.

Miller, James G. "Living Systems: Basic Concepts." *Behavioral Science*, July 1965,*10* (*3*), 193–237.

Mumford, Lewis. *Technics and Civilization*. New York: Harcourt, Brace and World, Inc., 1934.

Olmsted, Michael S. *The Small Group*. New York: Random House, 1967.

The Oxford English Dictionary. Oxford: Oxford University Press, 1971.

Parsons, Talcott. *Structure and Process in Modern Societies.* New York: The Free Press of Glencoe, 1960.

Porter, Lyman W. "A Study of Perceived Need Satisfactions in Bottom and Middle Management Jobs." *Journal of Applied Psychology*, 1961, *45*(*1*), 1–10.

Porter, Lyman W., and E. E. Lawler. *Managerial Attitudes and Performance.* Homewood, Illinois: Dorsey Press, 1968.

Rice, A. K. *Productivity and Social Organization: The Ahmedabad Experiment.* London: Tavistock Publications, 1958.

Roethlisberger, F. J., and W. J. Dickenson. *Management and the Worker.* Cambridge: Harvard University Press, 1939.

Shepard, J. *Automation and Alienation.* Cambridge: MIT Press, 1971.

Sommer, Robert. *Personal Space.* Englewood Cliffs: Prentice-Hall, Inc., 1969.

Sommerhoff, G. "The Abstract Characteristics of Living Systems." *Systems Thinking.* Ed. F. Emery. Baltimore: Penguin Books, Inc., 1969, pp. 147–202.

Srivastva, Suresh, Paul, Salipante, Jr., Thomas Cummings, William Notz; John Bigelow, and James Waters. *Job Satisfaction and Productivity.* Cleveland: Department of Organizational Behavior, 1975.

Steele, Fred I. *Physical Settings and Organization Development.* Reading: Addison-Wesley, 1973.

Tausky, Curt. "Meanings of Work: Marx, Maslow, and Steam Irons." Paper Presented at Annual Meetings of the American Sociological Association, August 1973.

Thompson, James D. *Organizations in Action.* New York: McGraw-Hill, 1967.

Thorsurd, Einer. "Industrial Democracy: Involvement, Commitment, Action: Some Observations During Field Research." Tavistock Institute of Human Relations, Doc. No. T. 886, 1966.

Trist, Eric L. "A Socio-Technical Critique of Scientific Management." A paper contributed to the Edinburgh Conference on the Impact of Science and Technology, Edinburgh University, May 1970.

Trist, Eric L., and K. W. Bamforth. "Some Social and Psychological Consequences of the Longwall Method of Coal Getting." *Human Relations*, 1951, *4*.

Trist, Eric L., and H. Murray. "Work Organization at the Coal Face: A Comparative Study of Mining Systems." Tavistock Institute of Human Relations, Doc. No. 506, 1958.

Trist, E. L.; G. W. Higgen, H. Murray, and A. B. Pollack. *Organizational Choice.* London: Tavistock Publications, 1963.

Vaill, Peter B. "Industrial Engineering and Socio-Technical Systems." *Journal of Industrial Engineering*, October 1967.

————. "Towards a Behavioral Description of High-Performing Systems." Prepared for a discussion at a colloquium of the School of Business Administration, University of Connecticut, 1973.

Van Bienum, H. J. J. "The Design of the New Radial Tyre Factory as an Open Socio-Technical System." Tavistock Institute of Industrial Relations, Doc. No. 150, 1968.

Van Beinum, H. J. J., and P. D. deBel. "Improving Attitudes Toward Work Especially by Participation." Tavistock Institute of Human Relations, Doc. No. HRC 101, 1968.

Vickers, G. *The Art of Judgment.* New York: Basic Books, 1965.

Vroom, V. H. *Work and Motivation.* New York: John Wiley and Sons, 1964.

Walton, Richard E. "The Diffusion of New Work Structures: Explaining Why Success Didn't Take." *Organizational Dynamics*, Winter 1975.

Weber, Max. *The Theory of Social and Economic Organization.* Trans. A. M. Henderson and Talcott Parsons and ed. Talcott Parsons. New York: The Free Press of Glencoe, 1947.

Wild, R. and T. Kempner. "Influence of Community and Plant Characteristics on Job Attitudes of Manual Workers." *Journal of Applied Psychology*, 1972, *56*(2), 106–113.

APPENDIX

ANALYTICAL MODELS FOR SOCIO-TECHNICAL SYSTEMS

The analytical models represent the work of researchers from the Tavistock Institute of Human Relations in London, England. We are indebted to Michael Foster, a major contributor to the models, for permission to reprint them in their original, working draft form. The first model, referred to as "analytical model for socio-technical systems," is intended for process or production units where some form of continuous technological process exists. The second model, titled "organizational objectives and role analysis," is an alternative method of analysis for work systems where no continuous process exists, such as service departments.

ANALYTICAL MODEL FOR SOCIO-TECHNICAL SYSTEMS

Step 1: Initial Scanning

The objectives of this step are to identify broadly the main characteristics of the production system and of the environment in which it exists and to determine, if possible, where the main problems lie and where the main emphasis of the analysis needs to be placed.

It is envisaged that the main method will be a briefing of the Action Group by the Departmental Manager or by someone deputed by him together with one or more discussion sessions.

The briefing needs to be fairly carefully structured and should cover the following ground:

a. The general geographical lay-out of the production system.
b. The existing organizational structure and the main groupings within it.
c. The main inputs into the system—with specifications where appropriate.
d. The main outputs from the system—again with specifications where appropriate.
e. The main transforming processes that take place within the system.
f. The main types of variance in the production system and their source, e.g., the main variances might arise from the nature of raw material or the nature of the equipment, or breakdowns, etc.
g. The main characteristics of the relationship between the production system and the department/refinery in which it exists.
h. The objectives of the System, both production and social.

Step 2: Identification of Unit Operations

The purpose of this stage is to identify the main phases in the production operation. Unit operations are here taken to be the main segments or phases in the series of operations which have to be carried out to convert the materials at the input end of the system into the products at the output end. Each unit operation is relatively self contained and each effects an identifiable transformation in the raw material; a transformation in this sense being either a change of state in the raw material, or a change of location or storage of the material.

The actions necessary to effect the transformation may be carried out by machines or by men but we are not at this stage concerned with either the characteristics or needs of the machines (e.g., maintenance needs, operating characteristics, etc.) or the characteristics and needs of the men (e.g., psychological needs). *The focus of attention is entirely on the series of transformations through which the raw material goes.*

Where possible the purpose of each unit operation needs to be identified in terms of: its inputs, its transformations and its outputs.

Step 3: Identification of Key Process Variances and Their Inter-Relationship

The objectives of this stage are to identify the key process variances and the inter-relationship between them.

By variance, we mean deviation from some standard or from some specification. It is necessary to emphasize two points:

a. At this stage in the analysis we are concerned with variance that arises from the raw material or from the nature of the process itself as it is currently or normally operating. We are not concerned with variance that arises from faults in the technical equipment or plant (e.g., breakdown or malfunctioning) nor are we concerned with variance that arises from the social system (e.g., maloperation or human error).
b. We are not concerned with the total range of variance. From studies of this type that have been carried out it has been found that in any production system there are a large number of variances that have either no effect or a comparatively minor effect on the ability of the production system to pursue its objectives. We are not concerned with variance of this order at this stage in the analysis, although it may be necessary to take some such variance into account in subsequent attempts to reach a higher level of joint optimization. At this stage we are concerned only with those variances which significantly affect the capability of the production system to pursue its objectives in one or more of its unit operations. We have called these variances "key variances" and propose certain criteria for their identification (see iii(b) below). The sequence of actions necessary to carry out this stage of the analysis is as follows:
   (i) *Identification of all variances in the system (arising from the nature of the raw material or from the nature of the process) which, in the opinion of the Action Group it is considered worthwhile to take note of.* The main source of information will be the manager and supervisors of the system, drawing upon their knowledge and experience. From experience gained in similar studies it will be necessary to work over this list of variances several times to ensure that all the main variances are included.
   (ii) *Drawing up a matrix of variances.* Taking the variances identified in (i) above, drawing up a matrix, the main purpose of which is to help show up any localized clusters of variances—control problems. It will also begin to show where information loops exist or are necessary in the production system. In addition it will also help in the selection of key variances (e.g., variances that have an effect through a series of unit operations are likely to be considered key to the control of the process).

(iii) *The identification of the key variances.* We propose that this be done in two stages:

(a) The department or unit manager and his appropriate assistants should make out a list of what they consider to be the key variances—drawing upon their experience and knowledge of the production system; and

(b) The Action Group should work over this list, checking it out against the matrix of variances and against the following four criteria. *A variance should be considered "key" if it significantly affects any one of these:*
—quantity of production
—quality of production
—operating costs (use of utilities, raw material, overtime, etc.)
—social costs (e.g., the stress, effort or hazard imposed on the men)

The first three dimensions are concerned with the *system's production objectives.* The last one is concerned with production system's social objectives, that derive from the social objectives written into the Company's Statement of Objectives and Philosophy. With the identification of these key variances it is possible to move into an analysis of the social system; to examine the way in which it contributes to their control, and so to the attainment of the production system's objectives, and also to examine the extent to which the social system's own needs are met.

Step 4: Analysis of the Social System

The objective of this step is to identify the main characteristics of the *existing social system.* It is not intended, however, to map or describe it in all its aspects, with all its complex sets of inter-relations and groupings, both formal and informal. It is hoped that by structuring the analysis carefully it will be possible for the analytical team—the Action Group—to draw out relatively quickly, sufficient of the relevant information to enable it to begin to develop job-design proposals. The following steps are, we consider the minimum necessary:

a. *Brief review* of the organizational structure where necessary filling in a little more detail than was included in Step 1 on number of levels, social groupings and types of roles.

b. *Table of variances control.* This is a key step in the analysis of the social system, *in terms of control of variance.* Its purpose is to show the extent to which key variances are presently controlled by the social system and it has been found possible by its use to identify where key organizational and informational loops exist or are required. It answers the following questions:
—Where in the process does the variance occur?
—Where is it observed?
—Where is it controlled?
—By whom?
—What tasks does he have to do to control it?
—What information does he get and from what source to enable him to carry out these control activities?

An additional column has been added to the table, headed "Hypotheses," because it has been found that in using the table suggestions or hypotheses for change tend to arise and it is considered worthwhile to note these at this stage for subsequent discussion and possible validation.

c. *Ancillary activities:* Filling out the descriptions of the workers' roles in the production system. In the variance control table mentioned above activities connected with the control of key variances will be listed. It is likely, however, that there will be a number of ancillary activities. Identifying these and trying to relate them to the control of the process may well lead, for example, to the identification of other forces operating in the social system, or the identification of additional key variances. On the other hand, it could conceivably lead to the questioning of these ancillary activities altogether.
d. *Spatial and temporal relationships.* Mapping out the physical or geographical relationship between the various roles in the production system (e.g., distances, or physical barriers between workers) and their relationship over time (e.g., over shifts or over a normal working day).
e. *Flexibility.* Using a mobility chart it is possible to identify the extent to which the workers in the production system share a knowledge of each others' roles. It may be necessary to carry out this step in two phases: an initial analysis, simply identifying where workers rotate, and a more detailed phase, where this appears appropriate, examining the extent to which they carry out the essential tasks associated with the roles. To be properly representative, the mobility chart should cover an adequate period, say two or three months. It is, therefore, considered appropriate to set up the recording of this information in the early weeks of the process of analysis.
f. *Payment system.* Setting out the pay relationship between various roles in the production system. This will have its impact of course upon job rotation, group working, etc.
g. *Psychological needs.* Testing out the roles against the list of psychological needs. It is considered that a simple adequate/inadequate rating against the main activities is sufficient. For this purpose the Action Group will need to rely on the manager's and his supervisor's or chargehand's perception of the roles. For the workers' perception of their roles, it will be necessary to set up some machinery for the collection of their views (cf. Step 5).
h. *Mal-operation:* identification of areas of frequent mal-operation and establish, where possible, causes.

Step 5: Men's Perception of Their Roles

This step is also part of the analysis of the social system. It is dealt with separately, partly because of its importance and partly because of the method of carrying it out. Its purpose is to learn as much as possible of the men's perception of their roles. *We are specifically concerned with the extent to which they see them fulfilling the psychological requirements.* It is considered that this should be accomplished perhaps by a personnel man attached to the Action Group, either as a full member or for this particular purpose.

It is proposed that he should run interviews with appropriate groups of men on two occasions: one within the first six weeks say of the analysis beginning, and the second towards the end of the process when job-design proposals are being finalized. Both interviews will be highly structured, designed by the Action Group with open-ended questions based on the general area of the psychological requirements, and in the case of the latter interview, on the developing jobs-design proposals.

With this step, the analysis of the production system itself is complete, and it is certainly to be expected that a number of redesign proposals or hypotheses will have emerged by this stage.

The analysis now goes on to consider the impact upon the production system of a number of "external" systems, e.g., maintenance, supply and user systems, refinery personnel policy, etc. These stages will influence any hypotheses that have emerged and may well throw up further redesign proposals.

Step 6: Maintenance System

This step is *not concerned* with the examination of the maintenance system or organization as such. It is concerned solely with the extent to which that system impacts upon the particular production system which is being analyzed. Its particular purpose is to identify the extent to which the maintenance system affects the capability of the production system to achieve its objectives. Thus, the objectives of this step may be stated as follows:

a. To determine the nature of the maintenance variance arising in the production system.
b. To determine the extent to which that variance is controlled.
c. To determine the extent to which maintenance tasks should be taken into account in the design of operating roles.

This does not mean, of course, that this analysis of maintenance variance is in any way subordinated to the analysis of process variance carried out in Step 3. Both are necessary to an understanding of the characteristics of the production system.

It may be in some cases that variance of a greater order arises from the maintenance system than from the production system itself, in which case one would expect greater emphasis upon this particular stage.

For the purpose of this stage we propose the collection of information on maintenance activities beginning within the first month of the project and being continued for say, two or three months. We consider that the collection of additional data and the burden of collection placed on operating and maintenance staff should be kept to the minimum consonant with achieving the objectives of the analysis, and the attached pro-formas are therefore, as simple as possible. They are, of course, in common with the rest of the model, open to development and improvement.

Step 7: Supply and User Systems

Once again this stage is *not concerned* with identifying the characteristic of the supply and user systems in themselves. The focus of the analysis is on the way in which these environmental systems affect the particular production system that is the focus of the project. The objectives of this stage are therefore:

a. To identify the variances that are passed into the production system *but that arise* in the system which supplies the raw materials, or the system which dispatches and (where appropriate) uses the products of the production system.
b. To examine, where this seems appropriate, the extent to which these variances could be controlled closer to their source, or their effect upon the production system diminished.

In general it is considered that the analysis across the boundaries of the production system should be kept to a fairly general level initially, and only continued in greater detail where there appears real possibility of effecting an improvement, e.g., a better control of variance or more appropriate flow of information.

The result of this step might either be a diminishing of the variance arising in the production system from across its boundaries, or in some cases, *a redefining of the production system's objectives* to ensure that they take realistically into account both supply and "marketing" constraints.

Step 8: Refinery Environment and Development Plans

The purpose of this stage is to identify those forces operating in the wider departmental or refinery environment that either affect the production system's ability to *achieve* its objectives, or which are likely to lead to a change in those objectives, in the foreseeable future. It has two main sub-steps:

a. *Development Plans*—The identification of any plans, either for the short or medium term or those long term plans which have a high probability of being implemented, for the development of the social or the technical systems. These clearly would have to be taken into account in the development of any redesign proposals.
b. *General Policies*—The identification of any general refinery policies or practices which impinge upon the production system, where these have not already been taken into account in the examination of the maintenance system and the supply/user systems. Examples of these might be the general method of promotion, which affects the social system, or the utilities supply and control system operating throughout the refinery, which affects the technical system.

Once again it should be emphasized that we are *not concerned* with an examination of the characteristics of these environmental systems as they exist in themselves, but only insofar as they affect the ability of the production system to pursue its objectives. In the analysis of most production systems these environmental factors will constitute "givens" rather than areas to be included in proposals for change.

Step 9: Proposals for Change

The purpose of this stage is to gather together all the hypotheses and proposals that have developed during the process of analysis, to consider their viability and to present them with sufficient structure to form the basis of a subsequent action program.

As has been mentioned above, it is likely that hypotheses will begin to arise as the analysis of the technical system is being completed. It is likely that these proposals will be added to, eliminated or modified as further information is gathered about the social system and above the environmental systems.

Those hypotheses that remain at the end of the process of analysis have to be tested out, as far as is possible on a theoretical basis, against appropriate criteria before being developed into viable proposals. The actual mix of criteria will vary from project to project and will require careful design. These criteria must, however, relate to the production systems objectives—they must cover, that is:

a. The production objectives of the system—concerned with production in terms of quantity, quality and general operating costs. This covers proposals specifically aimed at increasing the control over or diminishing variance in the production system.
b. The social objectives of the production system. This covers proposals aimed at, for example, increasing the extent to which psychological needs are met in role design and those aimed at diminishing the costs borne by the men in the social system (e.g., stress, hazard or heavy labour).

Many proposals will, of course, lie in both areas, e.g., proposals aimed at increasing the level of responsibility in the lower levels would both meet the psychological requirements and would perhaps lead to shorter lines of communication and more effective variance control. In addition, any proposals for the redesign of the social system must be tested out against emergency and crisis needs. In the case of a process unit for example this would entail the ability to shut the unit down in the event of loss of power or feed or major fire.

ORGANIZATIONAL OBJECTIVES AND ROLE ANALYSIS

Introduction

This model has been developed in conjunction with the socio-technical model as an alternative method of analysis for departments where no continuous process exists, such as service or advisory departments. Like the socio-technical model, its purpose is to help managers to analyze their existing organizations, as they currently and normally operate, and to produce proposals for change where this seems likely to lead to improved performance. The model is very much in the development stage and it is hoped that it will be improved by the contributions of members of the seminar. There are seven steps:

Step 1: General Scanning

This should provide a general introduction to the outputs, inputs, and transformation processes in the department, i.e., its objectives, its work and its organizational structure and location within the overall organization, as well as the geographical layout of the department. This scanning is necessary so that the more detailed investigations of these areas, which will follow later, can be seen against an overall background. One problem here is to decide upon the amount of detail at this level: in general this should be kept small. It is probably useful for the departmental manager, supposing this analysis is being undertaken by an Action Group, to describe what is going on rather than to explain its purpose, leaving the other members of the group to ask for reasons if they feel so moved.

Step 2: The Objectives of the System

It is important, if possible, to arrive at a clear definition of objectives, since this provides a rational datum against which to judge all activities in the department. In practice, the identification and statement of objectives pose difficulties since, for example:

a. The objectives stated may be so general as hardly to be a guide to action.
b. They may be multiple, but only one or two may be identified.
c. They may be non-measurable.
d. They may refer to several time periods, c.f., Philosophy Statement, Specific Objectives A and B.
e. They may be partly derived from a higher or other system levels.
f. They may be outputs which the system wants to minimize rather than to maximize, e.g., waste.
g. They may involve changes of the internal structure of the system rather than outputs from the system (i.e., change in assets).

In order to cope with these problems the following method of analysis is proposed:

To consider firstly all major *outputs* of the department whether they be processed raw materials, communications, man, or anything else. Secondly, to try and identify all inputs. Thirdly, to follow through these inputs and determine the steps by which they are processed before they become outputs; to make sure, in fact, that no significant, output has been missed. These outputs are then tested out to determine whether they are objectives by presenting them to the manager of the next higher system level, and asking him whether or not these are the outputs required. However, these are not the only input transformations: it is clear that a department is not only utilizing its inputs in order to process them and transform them into required outputs, but some inputs are coming into the department in order to maintain or develop the assets, and part of the objectives of a department will be directed towards these two activities. The assets of a department should be taken to include its plant and equipment, the money over which the manager may have authority, and the men.

In considering a department's outputs, a problem may arise in that it may not be possible to describe an output such as a communication (written or verbal) meaningfully unless some indication is given of its necessary contribution to an overall decision being made outside the boundaries of the department. In such cases it is useful to draw up a table with the following headings:

—Description of output,
—To whom sent,
—To what overall decision was it intended to contribute,
—What was the required contribution of the department's output to the overall decision,

and a final column indicating the consequences of sub-standard performance in respect of the social and economic cost, both inside and outside the department. The above analysis will determine the resources which are within the boundaries of the department, and those the manager needs to call upon. The departmental objectives should now be clearer and against this background it should be possible to begin to hypothesize the

a. responsibilities
b. authorities
c. information/communication links with others
d. key methods and procedures

which are appropriate, and to match these against those which exist and are identified particularly in Steps 3 and 4.

Step 3: Analysis of the Roles in the System

An analysis of each role in the system in the same way used in Step 2 in order to arrive at the role objectives, relating them back to the overall departmental objectives. This process should start at the top with the manager's role, and work downwards.

Step 4: Grouping of Roles

The analysis will identify the necessary role interaction links in so far as the current process exists and will lead to hypotheses about the clustering of these roles in respect of their geographical and temporal distribution, and status dimensions.

Step 5: Measurement of Roles Against Psychological Requirements

Having identified the inputs, transformations and outputs of each role, it is then useful to measure:

a. The manager's perception, and
b. The man's own perception of how far each role meets the psychological requirements as set out in the philosophy. The men's perception of their own roles can be achieved by individual interviewees to be carried out preferably by someone outside the department.

Step 6: Develop Change Proposals

It can be expected that in the course of the preceding steps various hypotheses for change will have emerged. These should now be refined into proposals for the redesign of jobs or organization—for example, a change in authorities or methods of grouping; or it may be that analysis by this stage will have indicated a need for a reformulation of departmental objectives. Proposals for change will, of course, have to be related to the wider environment of which a department is a part.

Step 7: Management by Objectives

Once the objectives of the department and its constituent roles have been determined, attention should be given to developing measurements of performance; the setting of targets (either jointly agreed with a manager or for self-monitoring), and how these results might be fedback to the man occupying the role (so that a man should know not only "what his job is (but) how he is performing in it" . . . Philosophy Statement).

Author Index

Subject Index